What SUCCESSFUL Math Teachers Do, Grades 6-12

To Barbara for her support, patience, and inspiration.
To my children and grandchildren: David, Lisa,
Danny, Max, and Sam, whose future is unbounded,
And in memory of my dear parents,
Alice and Ernest, who never lost faith in me.

—Alfred S. Posamentier

To my wife,
Tae-Jin,
who has made every sacrifice
to ensure my happiness and success.
To my children, Jennifer and Rebecca, who are an eternal
source of pride and joy, and in memory of my parents, Stanley
and Beatrice, who were always there for me.

—Daniel Jaye

What SUCCESSFUL Math Teachers Do, Grades 6-12

Alfred S. Posamentier • Daniel Jaye

For information:

Corwin Press
A Sage Publications Company
2455 Teller Road
Thousand Oaks, California 91320
www.corwinpress.com

Sage Publications Ltd.
1 Oliver's Yard
55 City Road
London EC1Y 1SP
United Kingdom

Sage Publications India Pvt. Ltd.
B-42, Panchsheel Enclave
Post Box 4109
New Delhi 110 017 India

Printed in the United States of America

Library of Congress Cataloging-in-Publication Data

Posamentier, Alfred S.
What successful math teachers do, grades 6–12: 79 research-based strategies for the standards-based classroom / by Alfred S. Posamentier and Daniel Jaye.
p. cm.
Includes bibliographical references and index.
ISBN 1-4129-1618-6 (cloth) — ISBN 1-4129-1619-4 (pbk.)
1. Mathematics—Study and teaching (Secondary)—Standards—United States.
I. Jaye, Daniel. II. Title.
QA13.P67 2006
510′.71′2—dc22 2005019975

This book is printed on acid-free paper.

05 06 07 08 09 10 9 8 7 6 5 4 3 2 1

Acquisitions Editor: Faye Zucker
Editorial Assistant: Gem Rabanera
Production Editor: Melanie Birdsall
Copyeditor: Kristin Bergstad
Typesetter: C&M Digitals (P) Ltd.
Proofreader: Liann Lech
Indexer: Sheila Bodell
Cover Designer: Michael Dubowe

Contents

Prologue

As a direct result of federal pressure on the states to continuously improve their instructional program and ensure that all students are being reached in the teaching process, teachers are being called on to meet professional standards and base their work on research-proven methods of teaching. Educational research, often conducted at universities or on educational sites by university researchers, is reported in educational journals and is most often read by other researchers. All too often the style in which research reports or articles on research findings are reported is not friendly or appealing to the classroom teacher. The very community—classroom teachers—that could benefit enormously from the findings of many of these educational initiatives rarely learns about these endeavors. It is the objective of this book to bring some of the more useful research findings to the classroom teacher. In our quest for the most salient research findings we were guided by the National Council of Teachers of Mathematics (NCTM) standards. Rather than merely presenting the research findings that support the standards, we have attempted to convert them into useful classroom strategies, thus capturing the essence of the findings and at the same time putting them into a meaningful context for the practicing mathematics teacher.

This book is to serve as a resource for mathematics teachers. It should provide a portal to access the many worthwhile findings resulting from educational, psychological, and sociological research studies done in Europe and in the United States. Heretofore, teachers have had very few proper vehicles for getting this information, short of combing through the tomes of research reports in the various disciplines. This book is designed to provide an easy way for the classroom teacher to benefit from the many ideas embedded in these academic exercises.

The book is designed to be an easy and ready reference for the mathematics teacher. It consists of six chapters, each with a theme representing one aspect of the typical instructional program. Each chapter presents a collection of teaching strategies, concisely presented in a friendly format:

The Strategy

This is a simple and crisp statement of the teaching strategy we recommend.

What the Research Says

This offers a discussion of the research project that led to the strategy. This section should simply give the teacher some confidence in, and a deeper understanding of, the principle being discussed as a "teaching strategy."

Teaching to the NCTM Standards

Here we present the salient NCTM standard that we are supporting with the strategy.

Classroom Applications

This section tells the teacher how the teaching strategy can be used in the mathematics instructional program. Where appropriate, some illustrative examples of the teaching strategy in the mathematics classroom are provided.

Precautions and Pitfalls

This is the concluding section for each strategy and mentions some of the cautions that should be considered when using this teaching strategy so that the teacher can avoid common difficulties before they occur, thereby achieving a reasonably flawless implementation of the teaching strategy.

The Sources

These are provided so that the reader may refer to the complete research study to discover the process and findings in greater detail.

We see this book as a first step in bringing educational research findings to the practitioners: the classroom teachers. Perhaps teachers will see that there is much to be gained to enhance teaching by reviewing educational research with an eye toward implementing the findings in their instructional program. Furthermore, it would be highly desirable for researchers to make more of an effort to extend their publications/findings to the classroom teacher. To do otherwise would make the entire activity of educational research irrelevant.

As you read the many instructional suggestions offered in this book we hope you will continuously think of yourself as the teacher who might implement them. Remember, your personality plays a large role in mapping out an overall instructional strategy; nevertheless, the specific research-based tips and strategies offered here will help you focus on certain aspects of your teaching. Teachers who continuously self-evaluate their instructional performance will, undoubtedly, become master teachers.

Acknowledgments

The authors wish to thank Noelle Jayne Lee for her assistance in performing some necessary editing functions in the preparation of the manuscript. Additionally, we acknowledge Kristin Bergstad's careful and concerned copyediting.

Publisher's Acknowledgments

Corwin Press thanks the following individuals for their contributions to this book:

Beverly R. Bryde, Associate Professor of Education, California Lutheran University, Thousand Oaks, CA

Charles A. Espalin, Director/Professor, USC School Counseling Program, Los Angeles, CA

Julia Green, National Board Certified Teacher, Pontiac, MI

Deborah Gordon, Teacher, Madison School District, Phoenix, AZ

Melissa Miller, Science Chair, Randall G. Lynch Middle School, Farmington, AR

M. Brad Patzer, Instructor, Mountain View High School, Idaho Digital Learning Academy, Medimont, ID

Kimberly C. Smith, Math Department Chair/Advanced Math Teacher, Welborn Middle School, High Point, NC

About the Authors

Alfred S. Posamentier is Dean of the School of Education and Professor of Mathematics Education of The City College of the City University of New York. He is the author and coauthor of more than 35 mathematics books for teachers, secondary school students, and the general readership. He is also a frequent commentator in newspapers on topics relating to education.

After completing his AB degree in mathematics at Hunter College of the City University of New York, he took a position as a teacher of mathematics at Theodore Roosevelt High School in the Bronx (New York), where he focused his attention on improving the students' problem-solving skills and at the same time enriching their instruction far beyond what the traditional textbooks offered. He also developed the school's first mathematics teams (both at the junior and senior level). He is currently involved in working with mathematics teachers and supervisors, nationally and internationally, to help them maximize their effectiveness.

Immediately upon joining the faculty of The City College (after having received his master's degree there), he began to develop inservice courses for secondary school mathematics teachers, including such special areas as recreational mathematics and problem solving in mathematics.

Dr. Posamentier received his PhD from Fordham University (New York) in mathematics education and has since extended his reputation in mathematics education to Europe. He has been visiting professor at several European universities in Austria, England, Germany, and Poland, most recently at the University of Vienna (Fulbright Professor in 1990) and at the Technical University of Vienna.

In 1989 he was awarded an Honorary Fellow at the South Bank University (London, England). In recognition of his outstanding teaching, The City College Alumni Association named him Educator of the Year in

1994, and New York City had May 1, 1994, named in his honor by the President of the New York City Council. In 1994, he was also awarded the Grand Medal of Honor by the Federal Republic of Austria. In 1999, upon approval of Parliament, the President of the Federal Republic of Austria awarded him the title of University Professor of Austria; in 2003, he was awarded the title of Ehrenbürger (Honorary Fellow) of the Vienna University of Technology, and he was recently (June 2004) awarded the Austrian Cross of Honor for Arts and Science, First Class by the President of the Federal Republic of Austria. In 2005, he was elected to the Hall of Fame of the Hunter College Alumni Association.

He has taken on numerous important leadership positions in mathematics education locally. He was a member of the New York State Education Commissioner's Blue Ribbon Panel on the Math A Regents Exams. He served on the Commissioner's Mathematics Standards Committee, which was charged with redefining the Standards for New York State, and he is on the New York City Public Schools Chancellor's Math Advisory Panel.

Now in his thirty-sixth year on the faculty of The City College of New York, he is still a leading commentator on educational issues and continues his long-time passion of seeking ways to make mathematics interesting to teachers (see *Math Wonders: To Inspire Teachers and Students,* 2003), students, and the general public—as can be seen from his latest books: *Math Charmers: Tantalizing Tidbits for the Mind* (2003); Π; *A Biography of the World's Most Mysterious Number* (2004); *101 + Great Ideas for Introducing Key Concepts in Mathematics, Second Edition* (2006); and *The Fabulous Fibonacci Numbers* (2006).

Daniel Jaye is the Assistant Principal for Mathematics at Stuyvesant High School in New York City. He lectures frequently and enjoys presenting interesting techniques in problem solving as well as problems that provide enrichment for the mathematics classroom. He is also interested in comparing math standards throughout the nation and the world.

Daniel Jaye graduated from the City College of New York with a major in mathematics and began his career teaching mathematics at Seward Park High School (New York City). After one year, he was invited to teach at the prestigious Stuyvesant High School (New York City), where he distinguished himself by teaching the entire range of high school mathematics courses.

After receiving his master's degree in mathematics education from The City College of New York, he took an interest in guiding student research projects in mathematics. Shortly thereafter he served as the math research coordinator and coordinated the submission of thousands of student-generated research papers to local and national competitions, including the Westinghouse and Intel Science Talent Search Competitions. In 2001 he was awarded the Mathematical Association of America's Edyth Sliffe

Award for Excellence in Teaching. He was also the recipient of Education Update's Outstanding Teacher of the Year award in 2004.

After twenty-five years of outstanding teaching, he was selected as Assistant Principal for the Stuyvesant High School Department of Mathematics. He immediately began to put his energies into creating more opportunities for talented and gifted students to study advanced mathematics. He was chosen to be the Executive Director of the New York City Math Team, where he coordinates the training activities of the 100 members of the team. In 2001 he created and directed the CCNY Summer Scholars Academy in Mathematics and Science. This program provides advanced courses in mathematics and is supplemented by a stellar guest lecture series, and features mathematics and science contests.

In 2004, he was chosen to serve on the New York State Math Standards Committee, which authored new state standards in mathematics. In 2004, he was elected President of the Association of Mathematics Assistant Principals for Supervision (New York City) and was awarded the Phi Delta Kappa Leadership in Education award. He currently serves on the New York City Public Schools Chancellor's Math Advisory Panel and the New York State Mathematics Curriculum Committee.

His passion for teaching and interest in mathematics standards were inspirational in creating this book.

1

Managing Your Classroom

STRATEGY 1: Create your own support network as soon as you begin your first teaching job.

What the Research Says

Teacher retention is a major concern of school districts nationwide. Inexperienced teachers are leaving the profession at an alarming rate, and they are being replaced by even less experienced teachers. This is creating a dangerous cycle that can have a profound effect on the learning community. Providing support for new teachers must be a priority of every district and school administration.

The Miami-Dade Public Schools have kept detailed records of school personnel and have conducted research on teacher retention. While they have found that turnover has significantly increased over the past decade, they did not attribute it to the salary scale. In fact, the research suggested that turnover was linked to "deterioration in the work environment with across-the-board policy changes that affect teachers."[1] It is generally accepted that inexperienced teachers are overwhelmed in their first year on the job. Though many contributing factors are cited, it is clear that lack of support is at the top of the list. The small school model typically has only a few math teachers, on staff. It is very difficult to obtain a balanced perspective on mathematics instruction from such a small group. The larger schools have many more teachers, but oftentimes are overcrowded. Teachers do not have common work areas, and there is not a feeling of community or a healthy collegial atmosphere. New teachers are left to fend for themselves. Simple tasks like making copies, obtaining textbooks, gaining access to technology, and obtaining supplies seem insurmountable. It is this culture and atmosphere that drives dedicated, well-intentioned individuals from the profession.

Teaching to the NCTM Standards

The NCTM Professional Standards addresses the development of the individual as a teacher of mathematics. Teachers should possess

> the growing sense of self as a teacher, and the continual inquisitiveness about new and better ways to teach and learn that serve teachers in their quest to understand and change the practice of teaching . . . they need to become the teacher of students under the guidance and support of both a cooperating practitioner and a mathematics educator.[2]

Classroom Applications

When you arrive at a school, your first goal should be to find a collaborative buddy—an experienced teacher who can be relied on to help you with some of the mundane aspects of your job. We have found that this is the single most important strategy to help new teachers survive their first year on the job. Newly licensed teachers are amazed at the hurdles they must face in their first year. Oftentimes, the training that they received in education programs did not adequately prepare them for the situation in which they now find themselves. Thus, having a buddy is essential. We suggest that new teachers use their buddies in many ways in addition to just procuring supplies. Collaborative buddies can share some aspects of the culture of the school. New teachers may wish to ask their buddies for some classroom management strategies, including how they take attendance, how they assign seats, how they assign homework, how they check homework, how often they give tests, what the format of their tests is, what additional resources are available for the students and teachers, whether there are class sets of calculators available, and so on. The best resource new teachers can have is the collective genius and experience of their colleagues.

Precautions and Possible Pitfalls

Remember that you have your own style of teaching and communicating. Although it is wonderful to solicit advice from veteran teachers, keep in mind that you have to adapt their suggestions to work with your particular strengths and weaknesses. Moreover, it is advisable to have many collaborative partners instead of just one. Then you can balance the ideas and suggestions and choose those that fit your personality and style.

Source

Burke, D., Cavalluzzo, L., Hansen, M., Harris, J., Lien, D., & Wenger, J. (2004, March). *Relative pay and teacher retention in Miami-Dade County public schools: Summary and research* (Report). Miami: Education Center Institute for Public Research.

STRATEGY 2: Before beginning a lesson, put an outline of what you are going to cover on the blackboard.

What the Research Says

Learning is more meaningful when students know in advance what is going to be covered in a lesson and how the teacher organizes the information to be learned. Seeing an outline on the board stimulates students' thinking about the various topics and helps them activate their prior knowledge about the topic. The connection between existing and new knowledge is an essential component of meaningful learning. The suspense or curiosity factor about the unknown is a useful device for motivation, and should be balanced with the prescription for the lesson.

Teaching to the NCTM Standards

The NCTM Learning Principle requires that students must "actively build new knowledge from experience and prior knowledge."[3] The research suggests that by providing an outline of what will be covered in class, students can begin to think about existing knowledge and concepts that are related to the day's lesson. For quite some time it has been standard practice to provide (or elicit) the "AIM" of the lesson and write it clearly on the board so that students will know the exact goal of the lesson. Providing an outline, as indicated above, takes that to a deeper level, providing an in-depth inventory of the concepts and skills that students will be learning. This can also be helpful to both the student and teacher to assess whether the stated goals of the lesson have been met.

Classroom Applications

There is a very clever method of teaching the theorems in geometry that deals with the measurement of an angle related to a circle. As a first step you might do a class on the various types of angles. Begin the lesson by putting an overview (advance organizer outline) of what you will be covering on the chalkboard. Example:

> Theorems in geometry involving various types of angles related to a circle:
>
> a. an inscribed angle
> b. an angle formed by two chords intersecting in the circle (not at the center)
> c. an angle formed by two secants intersecting outside the circle
> d. an angle formed by two tangents intersecting outside the circle
> e. an angle formed by a tangent and a secant intersecting outside the circle
> f. an angle formed by a chord and a tangent meeting at the point of tangency

This advance organizer outline will stimulate students to think about what they already know about angles (e.g., what types of angles they know of, such as a triangle; have they ever heard of an inscribed angle; are they familiar with the concepts of tangents, secants, etc.). It will also show students how you organize the ideas you are discussing—where one topic stops and another begins. You may want to include illustrations of each type of angle on the chalkboard so students can visualize the concepts and see whether or not they recognize them.

This activity is designed to demonstrate all of the measurements of the variations of an angle related to a circle. It can be carried out very nicely by cutting a circle out of a piece of cardboard and drawing a convenient inscribed angle on it. The measure of that angle should be the same as that formed by two pieces of string that are affixed to a rectangular piece of cardboard (see Figure 1.1).

Figure 1.1

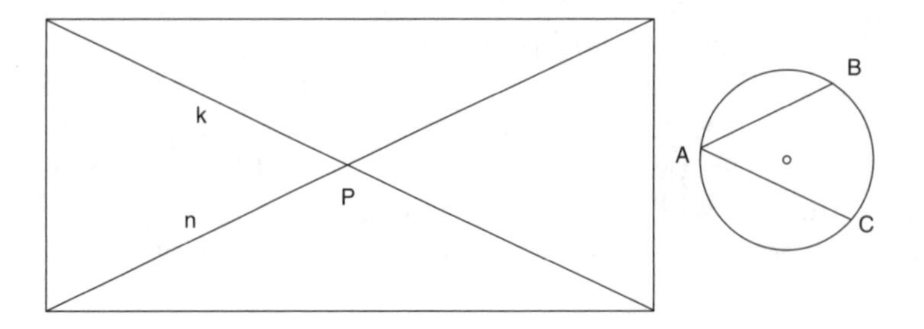

It is assumed that the theorem establishes that the measure of an inscribed angle of a circle is one-half the measure of the intercepted arc.

By moving the circle to various positions we will be able to find the measure of an angle formed by

- two chords intersecting inside the circle (but not at its center)
- two secants intersecting outside the circle
- two tangents intersecting outside the circle
- a secant and a tangent intersecting outside the circle
- a chord and a tangent intersecting on the circle

We begin with demonstrating the relationship between the arcs of the circle and the angle formed *by two chords intersecting inside the circle* (but not at its center). Position the cardboard circle so that A and C are on k, as in Figure 1.2.

Figure 1.2

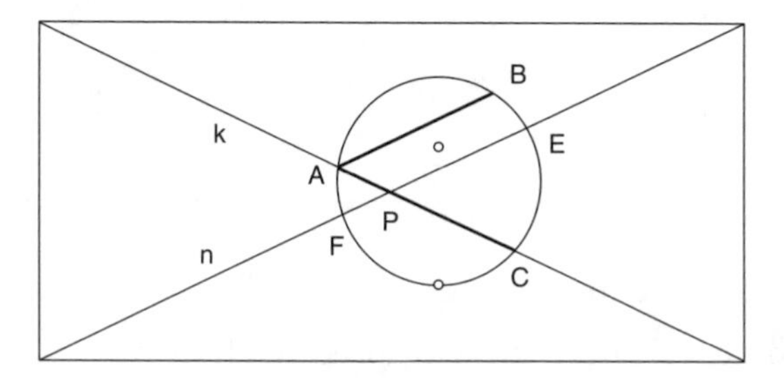

Notice that $m\angle A=\frac{1}{2}\widehat{BEC}$, and $m\angle A=m\angle EPC$. Therefore $m\angle P=\frac{1}{2}m\widehat{BEC}=\frac{1}{2}(m\widehat{BE}+m\widehat{EC})$. But, because parallel lines cut off congruent arcs on a given circle, $m\widehat{BE}=m\widehat{AF}$. It then follows that $m\angle p=\frac{1}{2}(m\widehat{AF}+m\widehat{EC})$, which shows the relationship of the angle formed by two chords, $\angle P$, and its intercepted arcs, $\widehat{AF}$ and $\widehat{EC}$.

Consider next the angle formed by *two secants intersecting outside the circle.* Place the cardboard circle into the position shown in Figure 1.3.

Figure 1.3

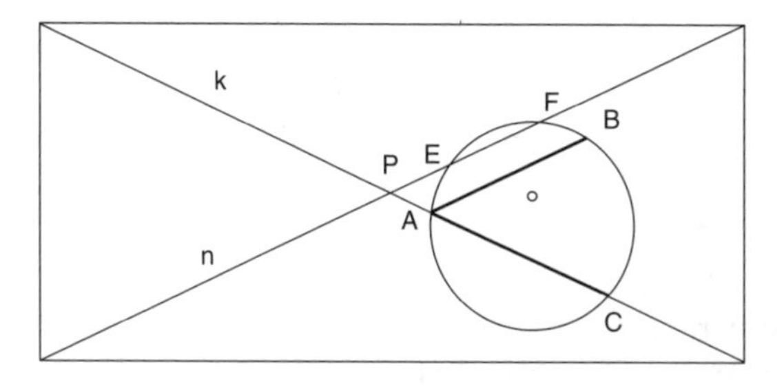

Begin by remembering that $m\angle P = \frac{1}{2}m\widehat{BC}$ and $m\angle FPC = m\angle A$. Because $m\widehat{AE} = m\widehat{BF}$, we can add it to and subtract it from the same quantity without changing the value of the original quantity. Thus,

$$m\angle P = \frac{1}{2}(m\widehat{BC} + m\widehat{BF} - m\widehat{AE} = \frac{1}{2}(m\widehat{FBC} + m\widehat{AE}).$$

In a similar way we can demonstrate the relationship between an angle formed by *two tangents intersecting outside the circle* and its intercepted arcs. We move the cardboard circle into the position shown in Figure 1.4.

Figure 1.4

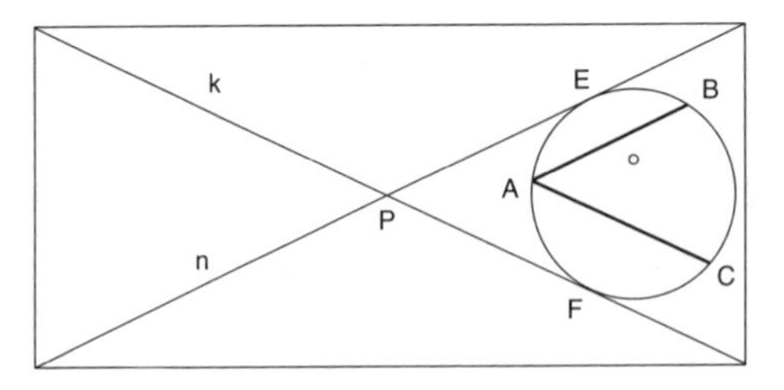

In this case, the equality of arcs, $\widehat{AE}$ and $\widehat{BE}$ as well as that of arcs $\widehat{AF}$ and $\widehat{CF}$ is key to demonstrating the desired relationship.

We have

$$m\angle P = m\angle A = \frac{1}{2}m\widehat{BC} = \rightleftharpoons \frac{1}{2}(m\widehat{BE} + m\widehat{BC} + m\widehat{CF} - m\widehat{AE} - m\widehat{AF})\swarrow\!\!\nearrow$$
$$= \swarrow\!\!\nearrow \frac{1}{2}(m\widehat{EBCF} + m\widehat{EAF}).$$

Again, by sliding the cardboard circle to the following position (see Figure 1.5) we can find the measure of the angle formed *by a tangent and a secant intersecting outside the circle.*

Figure 1.5

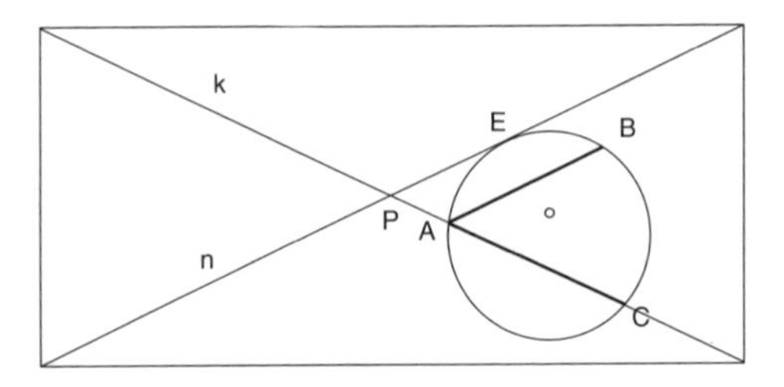

This time we rely on the equality of arcs. We get the following by adding and subtracting these equal arcs:

$$m\angle P = m\angle A = \frac{1}{2} m\widehat{BC} = \frac{1}{2}(mBC + m\widehat{BE} - m\widehat{AE}) = \frac{1}{2}(m\widehat{EBC} - m\widehat{AE}).$$

To complete the various possibilities of positions for the cardboard circle, place it so that we can find the relationship between *an angle formed by a chord and a tangent intersecting at the point of tangency* and its intercepted arc (see Figure 1.6).

Figure 1.6

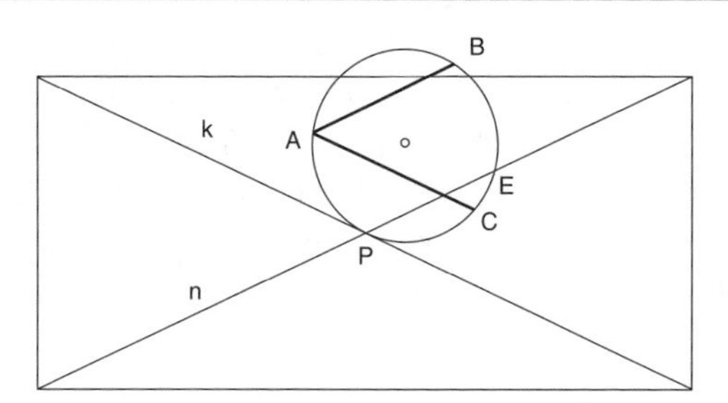

The crucial arc equality this time is $mAP = CP$, and $mAP = BE$. We begin as before:

$$m\angle P = m\angle A = \frac{1}{2} m\widehat{BEC} = \frac{1}{2}(m\widehat{BE} + m\widehat{EC} + m\widehat{PC} - m\widehat{AP})$$
$$= \frac{1}{2}(m\widehat{EC} + m\widehat{PC}) = \frac{1}{2} m\widehat{PCE}.$$

This activity can also be done quite nicely with a computer drawing program such as *Geometer's Sketchpad.*

Precautions and Possible Pitfalls

As mentioned above, this procedure for beginning a lesson has its advantages, but also the disadvantage of removing the "controlled surprise" factor from the lesson. It is important for the teacher to weigh this disadvantage against the gains when using this approach. Such professional judgments are always necessary when planning a lesson, but particularly in this case.

Source

Ausubel, D. (1960). The use of advance organizers in the learning and retention of meaningful verbal learning. *Journal of Educational Psychology, 51,* 267–272.

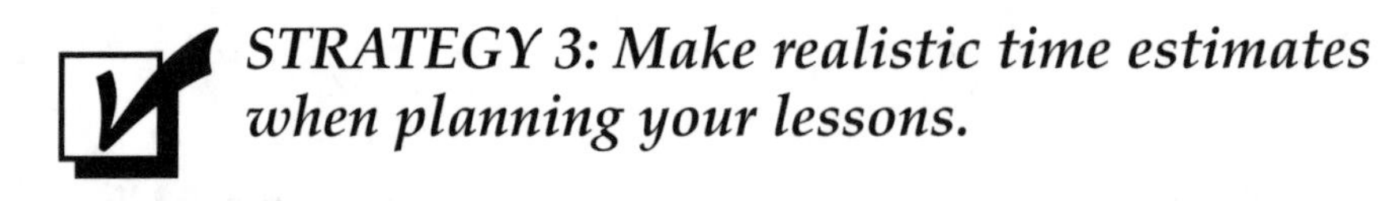

STRATEGY 3: Make realistic time estimates when planning your lessons.

What the Research Says

Teachers need to have excellent time management skills for students to learn effectively. It is sometimes said that "time + energy = learning." Sometimes teachers confuse time allocated for instruction with time students are actually engaged in learning. The concept of engaged time is often referred to as "time-on-task." Teachers often fail to take into account the time they end up devoting to managing student behavior and managing classroom activities. Teachers need to take this distinction between allocated and engaged time into account when estimating how much time it will take for students to learn a particular set of material. It's the time students actually spend learning that is the key to the amount of achievement.

Teaching to the NCTM Standards

In the NCTM *Research Companion to Principles and Standards for School Mathematics*, the subject of using time effectively in the classroom is addressed to ensure that lessons are planned to allow for students both to "learn to communicate mathematically and to communicate to learn mathematics. This requirement calls for planning that involves capitalizing on what students do and directing their activities toward important mathematics issues."[4] Effective planning is not simply choosing the right number of problems to support the lesson, but as the application below suggests, making important decisions about what to include in a given lesson so that meaningful learning takes place. "At the beginning of a discussion, the teacher might call on specific students selected in advance because he or she anticipates that a comparison of their solutions might lead to substantive mathematical conversation that advances the pedagogical agenda."[5] Allowing adequate time for student communication and involvement is key to planning a successful lesson.

Classroom Applications

Suppose you are planning a lesson on the introduction to the Law of Sines. You would like to develop or derive the Law, and you would like to have ample time to apply the Law to "practical" examples as well as the drill that typically follows the introduction of the Law. To fit this into a normal 50-minute lesson you might either relegate a more serious inspection of the derivation to a homework assignment and simply introduce the Law of Sines and its applications to the triangle, or you might search for a concise derivation of the Law, of Sines such as the following:

Consider the circumcircle of ΔABC with diameter $\overline{AD}$.

$$\text{Diameter AD} = \frac{AC}{\sin\angle ADC} = \frac{AC}{\sin\angle ABC}$$

$$\text{Diameter AD} = \frac{AB}{\sin\angle ADB} = \frac{AB}{\sin\angle ACB}$$

Figure 1.7

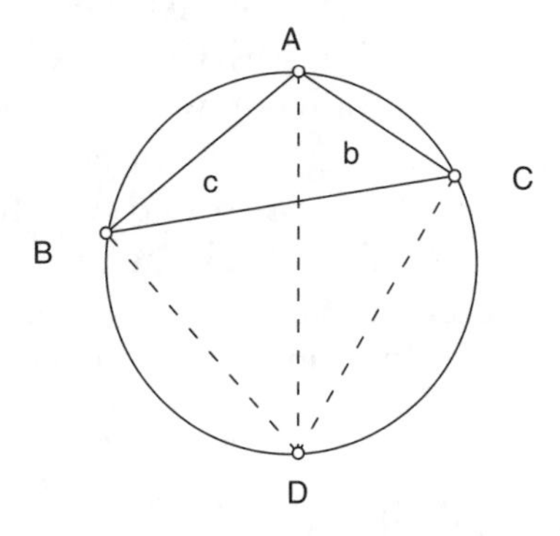

Therefore, $\frac{AB}{\sin\angle ADB} = \frac{AB}{\sin\angle ACB}$, which can then by followed in a similar way to the third part of the Law of Sines. This very concise proof will allow the teacher ample time to do a complete lesson on a topic which would otherwise require more than one lesson to introduce. In other words, it is wise for a teacher in planning a lesson and trying to ensure that there is enough time to do the lesson to also search for alternative methods which might be more concise and allow for a more streamlined lesson.

Precautions and Possible Pitfalls

Be sure to plan time in your "clever" lesson time for your students to digest the cleverness of proofs or demonstrations which may not be in the textbook. This will ensure that they have time to take complete notes of the development you are doing in the lesson so that review at home is then possible.

Source

Brophy, J. (1988). Research linking teacher behavior to student achievement: Potential implications for instruction of chapter 1 students. *Educational Psychologist, 23*, 235-286.

STRATEGY 4: Make classroom activities flow smoothly.

What the Research Says

Kounin's classic study of classroom management compared effective with ineffective teachers. The effective teachers' classes did not have many problems, while continuous disruption and chaos characterized the ineffective teachers' classes. By observing effective and ineffective teachers' classes, Kounin discovered that the major difference between them was in preventing problems rather than handling problems once they arose. One way teachers prevented problems was by making sure students were not sitting around waiting for the next activity, but were engaged in meaningful work all the time. Effective teachers had smooth transitions between activities, they conducted activities at a flexible and reasonable pace, and their lessons involved a variety of activities.

Teaching to the NCTM Standards

The NCTM Professional Standards addresses the effective learning environment for mathematics instruction. The standard includes "providing and structuring the time necessary to explore sound mathematics and grapple with significant ideas and problems; using the physical space and materials in ways that facilitate students' learning of mathematics."[6] Time is the most precious resource a teacher has in delivering mathematics instruction. Often the difference between a satisfactory lesson and an outstanding lesson is the one or two "extra" applications that the students can explore during a lesson. These applications are typically more challenging and their inclusion is possible because of outstanding time management by the instructor. The application below gives one method of maximizing the use of time to review the homework assignment from the previous day's lesson. A skillfully prepared lesson should be artfully choreographed to provide a seamless transition from one activity to the next.

Classroom Applicationss

The smooth beginning of a class is quite obviously an important aspect of an effective lesson. Rather than spending time on having students copy their homework on the chalkboard at the beginning of a lesson, a smooth beginning would have them immediately bring their work to the class by showing it on an overhead projector (or video monitor). Students should have their model solutions on a transparency when they enter the classroom (done the evening before as part of their homework) so that they can present their work without taking class time to copy it onto a transparency.

Another example of a smooth transition is for the teacher to structure the order of topics to be presented in a logical way so that one comfortably leads into the next, or so that one situation is an elaboration of its predecessor, and so on. Teachers should make sure transitions are smooth both in terms of topics presented (content) and student activities (process). With the teacher conscious of these transitions, the likelihood of them happening is greatly increased.

Precautions and Possible Pitfalls

Be aware of the smoothness of a lesson's start. Student behavior and overall attitude will determine the appropriateness of a lesson's level and tone. Be prepared to make modifications and adjustments to the students' style (within reason, of course), and remember, there is not one best method. The "best method" may vary from class to class!

Some general suggestions follow:

1. Avoid interrupting students while they are busy working.
2. Avoid returning to activities that have been treated as finished.
3. Avoid starting a new activity before finishing a preceding activity.
4. Avoid taking too much time when beginning a new activity. Slowing down can breed trouble.

Source

Kounin, J. (1970). *Discipline and group management in classrooms.* New York: Holt, Rinehart & Winston.

STRATEGY 5: Have "eyes in the back of your head" so you notice misbehavior at an early stage.

What the Research Says

In Kounin's classic study of classroom management, the major difference between effective and ineffective teachers' classrooms was in preventing problems rather than handling problems after they arose, thus causing continuous disruption and chaos. By observing effective and ineffective teachers' classes, Kounin discovered that one way teachers prevented problems was by "with-it-ness." This was the teachers' ability to be aware of what individual students were doing at all times. They were not missing what was going on around the classroom while they were working with a few individual students. These teachers were described as having "eyes in the back of their heads." With this alertness, teachers were able to prevent minor problems from becoming major problems and, when a problem did arise, they were able to pinpoint the cause or culprit and put a quick end to the disturbance. This enabled the teacher to deal expeditiously with the problem at hand and effectively avoid its recurrence.

Teaching to the NCTM Standards

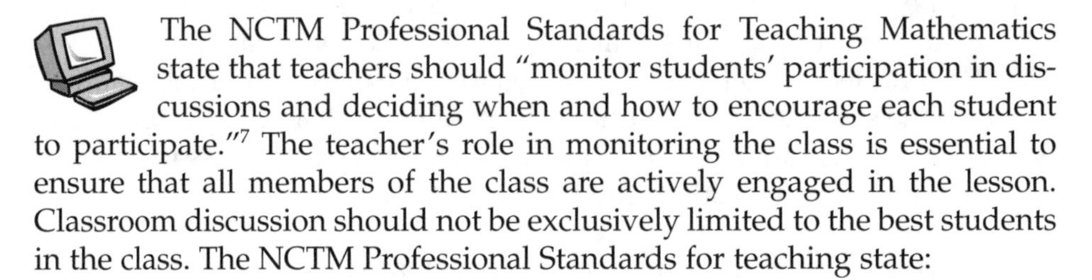

The NCTM Professional Standards for Teaching Mathematics state that teachers should "monitor students' participation in discussions and deciding when and how to encourage each student to participate."[7] The teacher's role in monitoring the class is essential to ensure that all members of the class are actively engaged in the lesson. Classroom discussion should not be exclusively limited to the best students in the class. The NCTM Professional Standards for teaching state:

> It is a key function of the teacher to develop and nurture students' abilities to learn with and from others. . . . Classroom structures that can encourage and support this collaboration are varied: students may at times work independently, conferring with others as necessary; at other times students may work in pairs or small groups.[8]

The applications and suggestions below provide a variety of "tricks," techniques, and methods by which the teacher can keep the entire class on task during a lesson. Engaging all of the members of the class is a proactive measure that supports teaching and learning and prevents students from misbehaving in class.

Classroom Applications

There are numerous "tricks" that teachers can use to move them in the right direction with respect to this sort of with-it-ness:

- Remember the spot from which you address the class is the temporary "front of the room." Constantly move around the classroom;

don't stand in one spot. This allows most of the class to be at the "front of the room" at least for a part of the lesson.

- Talk to the class even as you write. This will remind them that your thoughts are still on them and not on the board-work. Don't turn your back to the class, except for very brief intervals.
- Don't talk to one student while ignoring the rest of the class. If you speak to one student, do so while looking at the rest of the class at the same time.
- Don't always call on volunteer responses to teacher questions. Call on those students who tend to look away from the teacher as soon as a question is asked.
- Frequently make eye contact with students who tend to cause disruptions so they know you're paying attention to them.
- While working with one cooperative learning group, look around the classroom to see how other groups are working. Make comments as necessary. Let them know that although you are working with one group, you are not ignoring what the other groups are doing.

Precautions and Possible Pitfalls

The way to be "with it" is an individual phenomenon. A teacher must do that which is consistent with his or her personality. To do otherwise would be counterproductive because students are very adept at picking up unnatural behavior. Bear in mind that there are no perfect universal solutions to maintaining proper class control. The above are merely suggestions that ought to be modified to meet individual personalities and strengths.

Source

Kounin, J. (1970). *Discipline and group management in classrooms.* New York: Holt, Rinehart & Winston.

STRATEGY 6: Help students develop self-control to enhance their thinking and independence, as well as to ease your own workload.

What the Research Says

Just a few small changes in your methodology could provide an increase in students' self-control. The increase of students' self-control does not release the teacher from his or her outside control. Outside control by the teacher is the basic prerequisite for step-by-step cultivation of self-control. Gradually transfer your control and guidance to students as they develop their own control and feelings of responsibility. In a study about students' self-control, students reported that their lack of self-control makes them feel uncertain about whether they really reached an educational objective. That's why students want external control by the teacher, and that's precisely why teachers need to guide students in realizing self-control.

The research has shown that:

1. High-achieving students have better self-control than students who have learning weaknesses. However, good students sometimes think that they do not need to practice self-control.
2. Preplanned self-control is hard to observe in students. When self-control is observed, it tends to be more reactive than proactive.
3. The more students are proactive in their self-control, the better they are in reacting with self-control.
4. Girls show a stronger tendency toward self-control than boys. Boys tend to skip steps of self-control or do it superficially.
5. Within situations of high demands (such as tests), students realize a greater degree of self-control. Self-control during homework tends to be considered superfluous.
6. When teachers control students' behavior, students tend to adapt to it and refrain from self-control.
7. Teachers' efforts to encourage students' self-control focus only on reactive or result-related self-control.
8. Students know only about techniques of result-related or reactive self-control. These include self-checking, using reference books, using calculators, or verifying the results of a calculation. Low-achieving students tended to mention verifying a calculation as the technique of self-control. Only a few high-achieving students identified making rough estimates as a technique of proactive self-control.

Teaching to the NCTM Standards

The NCTM Principles and Standards for School Mathematics state that, "A major goal of school mathematics programs is to create autonomous learners."[9] Giving students the responsibility for their own self-control is an important step toward allowing students to grow into independent learners. "Students learn more and learn better when they can take control of their learning by defining their goals and monitoring their progress."[10] "Effective learners recognize the importance of reflecting on their thinking and learning from their mistakes."[11] All of the applications below support this theme and advance the goal of creating autonomous learners.

Classroom Applications

Practice continuous methods of self-control like:

- Making rough estimates (Do not trust blindly in the calculator!)
- Using mathematical theorems
- Procedures for drawing representations
- Procedures for graphic representations
- Using templates and measuring instruments

Practical ways for step-by-step improvements in students' self-control consist of the following:

- Students make mutual comparisons of their answers and solutions strategies.
- When one student presents his or her way of solving a problem, another student should give feedback to the problem solver.
- Combine assignments with elements of playful self-control. This is suitable particularly for students with learning weaknesses or students with impulsive work habits.

At the end of this strategy you will find an example of an exercise with such a combination between assignment and playful self-control.

- Give assignments that force students to engage in proactive self-control:
  - Design a task that contains superfluous information.
  - Assign a problem or task that is not solvable or is solvable only under certain conditions.

With each of these techniques of self-control, a teacher is likely to complain about loss of time. However, the students' developing self-control abilities will save time in the long run.

Precautions and Possible Pitfalls

Strict demands on students to use techniques of self-control incessantly or indiscriminately can backfire. Students who already understand an algorithmic procedure will view the demanded checking as only a mechanical (and therefore meaningless) activity. As a result of this, these students may devalue the self-control.

Source

Frank Fischer & Kittlaus Bernd. Ergebnisse von Untersuchungen zur Selbstkontrolle bei Schülern der 7. Klasse im Mathematikunterricht (Results of investigations about self-control of 7th grade students in math classes). *Mathematik in der Schule,* 29 Jg., Heft 11 (1991) S. 761–768.

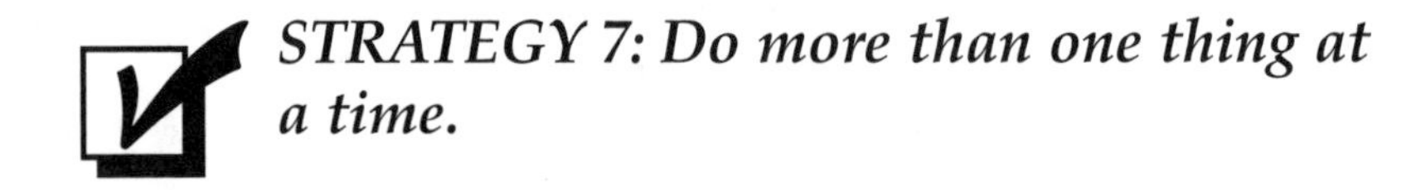

What the Research Says

Kounin's classic study of classroom management compared effective with ineffective teachers. The effective teachers' classes did not have many problems while the ineffective teachers' classes were characterized by continuous disruption and chaos. By observing effective and ineffective teachers' classes, Kounin discovered that the major difference between them was in preventing problems rather than handling problems once they arose. One way teachers prevented problems was by "overlapping" activities, or supervising and keeping track of several activities at a time. In order to successfully overlap activities, effective teachers continuously monitored what was going on in the classroom.

Teaching to the NCTM Standards

The NCTM Professional Standards for Teaching and Learning ask teachers to reflect upon the following question: "How well are the tasks, discourse, and environment working to foster the development of students' mathematical literacy and power?"[12] The effective mathematics teacher recognizes the varying ability levels of the students in the class and must adapt to provide a differentiated instructional model to challenge and engage all of the members of the class and keep them on task for the duration of the period. "Teachers must monitor classroom life using a variety of strategies and focusing on a broad array of dimensions of mathematical competence."[13]

Classroom Applications

There is an adage in the teaching world that the teacher needs to move continuously about the classroom and be omnipresent. A critical time for teachers to make their presence known is at the beginning of a lesson. While students are putting homework problems on an overhead transparency or on the blackboard, the teacher is free to work with individual students. While a teacher or student is collecting homework assignments, the teacher can be introducing the class to the next topic by posing a problem or question to tap and review students' prior knowledge of the topic they will discuss next. While walking around the classroom discussing a topic, a teacher can glance at students' desks to check for homework or to make sure students are looking at the appropriate material. Teachers can also discuss how to solve a problem while walking around and showing their presence to make sure students do not misbehave.

Precautions and Possible Pitfalls

The key thing to bear in mind is not to spread yourself too thin when assuming more than one responsibility at a time. In addition, don't move around the classroom so much that the movement becomes a distraction for students.

Source

Kounin, J. (1970). *Discipline and group management in classrooms.* New York: Holt, Rinehart & Winston.

STRATEGY 8: Work directly with individual students as often as possible.

What the Research Says

Frequent contact between teachers and students helps students develop academically and intellectually. Rich teacher-student interaction creates a stimulating environment, encourages students to explore ideas and approaches, and allows teachers to guide or mentor individual students according to their individual needs.

Teaching to the NCTM Standards

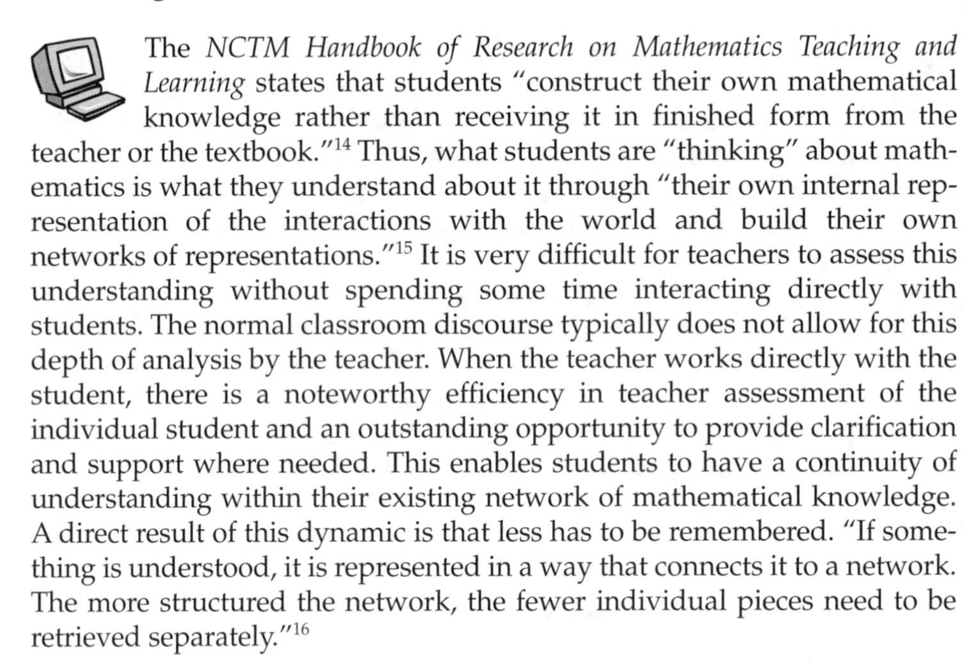

The *NCTM Handbook of Research on Mathematics Teaching and Learning* states that students "construct their own mathematical knowledge rather than receiving it in finished form from the teacher or the textbook."[14] Thus, what students are "thinking" about mathematics is what they understand about it through "their own internal representation of the interactions with the world and build their own networks of representations."[15] It is very difficult for teachers to assess this understanding without spending some time interacting directly with students. The normal classroom discourse typically does not allow for this depth of analysis by the teacher. When the teacher works directly with the student, there is a noteworthy efficiency in teacher assessment of the individual student and an outstanding opportunity to provide clarification and support where needed. This enables students to have a continuity of understanding within their existing network of mathematical knowledge. A direct result of this dynamic is that less has to be remembered. "If something is understood, it is represented in a way that connects it to a network. The more structured the network, the fewer individual pieces need to be retrieved separately."[16]

Classroom Applications

Working with individual students in a traditional classroom setting is not practical for long periods of time. While students are working individually on an exercise, the teacher should visit with individual students and offer them some meaningful suggestions. Such suggestions might include hints on moving a student who appears frustrated or bogged down on a point toward a solution. These private comments to students might also be in the form of advice regarding the form of the student's work. That is, some students are "their own worst enemy" when they are doing a geometry problem and working with a diagram that is either so small they cannot do anything worthwhile with it, or is so inaccurately drawn that it proves to be relatively useless. Such small support offerings will move students along and give them that very important feeling of teacher interest.

In some cases, when a student experiences more severe problems, the teacher might be wise to work with the individual student after classroom hours. In the latter situation, it would be advisable to have the student describe his work as it is being done, trying to justify his procedure and explain concepts. During such one-on-one tutoring sessions, the teacher can get a good insight into the student's problems. Are they conceptual? Has the student missed understanding an algorithm? Does he have perceptual difficulties? Spatial difficulties? And so on?

Precautions and Possible Pitfalls

To work with individual students and merely make perfunctory comments when more might be expected, could be useless when considering that the severity of a possible problem might warrant more attention. Teachers should make every effort to give proper attention to students when attempting to react to this teaching strategy. They should keep the student's level in mind so that where appropriate, teachers can add some spice to the individual sessions by providing carefully selected challenges to the student so that there may be a further individualization in the learning process. Make sure good students don't get bored. Challenge them by giving them more difficult problems to solve, having them tutor other students, or having them evaluate alternative approaches to solving a problem.

Source

Pressley, M., & McCormick, C. (1995). *Advanced educational psychology.* New York: HarperCollins.

STRATEGY 9: Use classwide peer tutoring to help your students learn, whether or not they have learning disabilities.

What the Research Says

A whole classroom of students helping other students has been found to be an efficient and effective method of enhancing achievement. Twenty teachers participated in a study of classwide peer tutoring with forty classrooms in elementary and middle schools. Half of the schools implemented classwide peer tutoring programs and half did not. Both urban and suburban schools participated in the study. Students came from diverse backgrounds, both culturally and linguistically. There were three different categories of students: average achievers, low achievers without learning disabilities, and low achievers with learning disabilities.

The peer tutoring programs were conducted three days a week, for thirty-five minutes a day, for fifteen weeks. Stronger students were paired with weaker students. Teachers reviewed each pair to ensure they were socially compatible. In all pairs, students took turns serving in the roles of tutor and tutee. Student pairs worked together for four weeks; then teachers arranged new pairings. Teachers received training on how to train their students to be tutors. Tutor training included teaching students how to correct each other's errors. Achievement tests were administered before and after the peer tutoring program. Regardless of whether students were average achievers or low achievers with or without learning disabilities, students in the peer tutoring classrooms achieved at higher levels than those in the classrooms without classwide peer tutoring.

Teaching to the NCTM Standards

The NCTM Professional Standards encourage and expect that students

> work independently or collaboratively to make sense of mathematics. . . . Students' learning of mathematics is enhanced in a learning environment that is built as a community of people collaborating to make sense of mathematical ideas. It is a key function of the teacher to develop and nurture students' abilities to learn with and from others.[17]

Powerful learning communities employ peer tutoring as an important component of teaching and learning. Empowering students to help one another advances the goal of having students become autonomous learners.

Classroom Applications

There are many areas in mathematics that lend themselves to a peer tutoring program. When there is a skill to be learned and all that one needs is experience with success (i.e., drill with immediate feedback), then peer tutoring could provide an efficient way to monitor and support a student trying to master the skill. Say a student has difficulty with factoring, and part of his problem is recognizing which type of factoring is called for. To compound the problem, more than one type of factoring may be used, making this doubly confusing. Here a peer tutor (under the guidance of a teacher) can be quite beneficial. A student who has difficulty doing geometric-theorem proofs could find that a peer tutor is a genuine asset. In addition, the tutor, in explaining the proof to the student, is also provided with an opportunity to strengthen his or her own understanding of the concept of proof (a higher-order thinking skill) and with the role of proof in mathematics. Thus, there is often mutual benefit in a peer tutoring program.

Precautions and Possible Pitfalls

A tutor training program offered by the teacher must precede peer tutoring. Tutors must be given some instruction on how to conduct the sessions, what sort of difficulties to look for on the part of the tutee, and what points to stress in the sessions (based on the teacher's assessment of the class). Any individual difficulties on the part of the tutees should be mentioned to the tutor prior to the sessions. Tutors should be taught to guide student learning, and *not* merely solve problems for students. Students with severe learning disabilities may be too disruptive to benefit from classwide peer tutoring unless the tutors first receive individualized instruction from learning disabilities specialists.

Source

Fuchs, D., Fuchs, L., Mathes, P. G., & Simmons, D. (1997). Peer-assisted learning strategies: Making classrooms more responsive to diversity. *American Educational Research Journal, 34*(1), 174–206.

STRATEGY 10: Encourage students to be mentally active while reading their textbooks.

What the Research Says

When comparing students who are good at understanding what they read with students who are poor at understanding what they read, research shows that good comprehenders are more mentally active than poor comprehenders. Mental activities that characterize good comprehenders include skimming, self-questioning, rereading, inferring, and visualizing. In addition to using such strategies for actively processing the text, good readers tend to coordinate their reading strategies to achieve comprehension.

Teaching to the NCTM Standards

Teachers of mathematics rely on their textbooks in their day-to-day teaching. Decisions on what to teach and how to teach it are informed by the choice of textbook that students are using. Educators must teach students how to use the textbook effectively. One of the primary functions of the textbook is to provide exercises for students to solve, as well as a variety of examples that highlight methods of solution. The approach that a particular author takes to the teaching and learning of mathematics influences the scope and sequence of the topics. In addition, the manner in which various process strands are incorporated into the content strands further magnifies the importance of choosing an appropriate text and using it correctly. Because the NCTM values both process and content, it is essential that teachers instruct students to use the textbook to compare and contrast solutions to a wide variety of problems. Students should not view the textbook simply as a source of problems, but rather as a source for seeking solutions. Thus, actively engaging a student with a good textbook provides the student with a safety net and affords the student the opportunity to take academic risks in problem solving. Where appropriate, sample exercises should be used to generate discourse on alternative solutions that are available.

Classroom Applications

It is unfortunate that most students do not use their mathematics textbook in the way authors would like them to be used. Typically, students only use textbooks to complete homework assignments or to prepare for a test. This use shortchanges many students, for the textbook could very well (and often does) provide alternative explanations to a concept explained by the teacher in class. Students would be well advised (and should even be urged) to read the explanatory material covered in class, for it is quite conceivable that a student's notes are not always complete or truly reliable. Reading a mathematics book is clearly not like reading a novel. The teacher ought to take time out from the normal mathematics instructional program to focus on the way a mathematics textbook ought to be read. By taking a small snippet of time from each of a series of lessons in order to consider the textbook and how it should be used, the teacher will be making the review (via the textbook) of future topics studied much more effective. The teacher should explain the notation and style of the author and indicate the author's pedagogical intentions and any other peculiarities that may be appropriate. Teachers should also help students develop the analytical skills for identifying when, why, and how a particular model described in the text fits a particular problem. Students must constantly question their understanding of each idea and look toward the overriding direction or "big picture" of the concepts or unit being developed and how they are related to other concepts. Oftentimes, mathematics textbooks offer model solutions to problems. These should also be read in a very active fashion before doing the exercises, even if the student thinks he or she can "fly" through the exercises after or without reading the explanatory material.

Precautions and Possible Pitfalls

The teacher should make a special point of instructing students to read the textbook regularly. In doing so, teachers should highlight specific aspects of the readings, such as the differences between class instruction and the textbook material (if such exists). The teacher should be aware that there may exist individual reading problems with students in the class that may not manifest themselves in their mathematics achievement. That is, a good mathematics student could be a poor reader. The teacher's awareness of and sensitivity to these weaknesses are important when considering the task of reading mathematics textbooks.

Source

Long, J. D., & Long, E. W. (1987). Enhancing student achievement through metacomprehension training. *Journal of Developmental Education, 11*(1).

STRATEGY 11: Avoid reacting emotionally when evaluating problematic situations in the classroom.

What the Research Says

An emotional reaction can prevent a teacher from objectively assessing a problematic situation. When a teacher displays a high level of emotional excitement, he or she tends to evaluate situations more negatively than is objectively appropriate. This reaction is especially common when the teacher is very sensitive to the perception that students are not reacting to the teacher's demand and when the teacher is very sensitive to the perception of the students' motor activity and body language. A cohort of 132 teachers participated in an investigation of how teachers evaluate problematic situations. The participants saw several videotapes with one-and-a-half- to four-minute sequences that represented problematic situations. After every scene teachers had to complete a questionnaire that asked for their observations, evaluation, estimation of the situation, emotion, and reaction. The extent of their emotional excitement was also requested.

Teaching to the NCTM Standards

The NCTM Professional Standards state that "the teacher of mathematics should create a healthy learning environment that fosters the development of students' mathematical power."[18] Although standards in mathematics do not specifically address behavioral issues, it is important to recognize that the comprehension and mastery of mathematics can sometimes be frustrating for the struggling student. Teachers' sensitivity to this issue can promote an atmosphere where student frustration can be channeled into productive activities that can put the student back on track. The application below highlights how a student's chronic motor activity may indicate an underlying problem in comprehension and not a behavioral problem. Teachers who are sensitive to this can provide immediate support to alleviate the frustration that accompanies student failure to comprehend mathematical concepts.

Classroom Applications

Beware of vicious cycles! Teachers frequently penalize students who show a high degree of motor activity or conspicuous body language. Students with conspicuous motor activity or pronounced body language may not be able to stop it immediately. Chronic motor activity or a student's present mood may prevent that student from controlling motor activity entirely. When a teacher gets the impression that students are not reacting to the teacher's demand to stop moving, teachers tend to evaluate the situations more negatively.

How to break the vicious cycle:

- Sanction the student.
- Be patient for a few seconds; students need time to realize just what you are asking of them. (Careful! These seconds can seem like ages from your perspective.)
- Establish eye contact. In this way the student will get the impression that you mean it.
- If it is necessary to repeat your demand, do it word for word, as originally presented. Do not confront the student with what might appear to be a different demand or request. This could confuse the issue.

Precautions and Possible Pitfalls

In giving the appearance of being simultaneously well balanced and strict, you have to work with high self-control. If you change suddenly and become very emotional, the student's behavior is likely to worsen. By maintaining your equilibrium—not reacting emotionally in problematic situations—you can prevent the cycle from starting again.

Source

Albert Thienel. Der Einfluß der emotionalen Betroffenheit von Lehrern auf das differentielle Erleben einer Problemsituation (The influence of emotional excitement of teachers by differential definitions of a problematic). *Psychologie in Erziehung und Unterricht,* 36 Jg., (1989) S. 210–215.

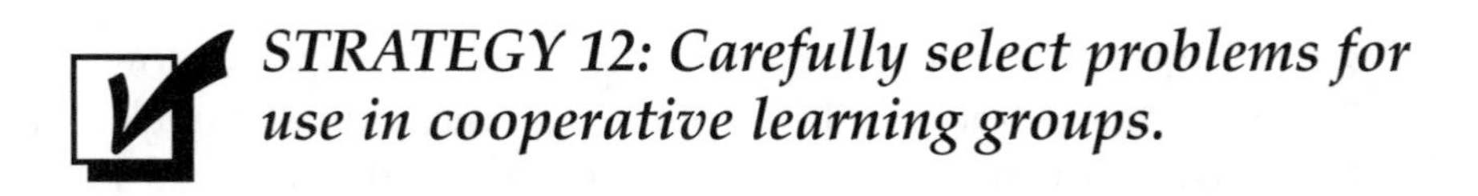

STRATEGY 12: Carefully select problems for use in cooperative learning groups.

What the Research Says

Cooperative learning in high school mathematics classes is often viewed by both teachers and students as having a beneficial effect on the learning environment. Both preferred small group work in comparison to traditional methods of instruction. Students became aware that the keys to learning are teaching and explaining. One study examined two teachers who used cooperative learning frequently in class; it investigated their and their students' perceptions about cooperative learning practices in a high school mathematics classroom. The researchers observed classroom activities and interviewed teachers and students. The teachers perceived cooperative learning as an asset to the learning environment, despite some of its limitations. The results showed that some teachers thought that cooperative learning could be used effectively with all math problems. However, most research indicates that careful selection of problems for use in cooperative learning groups is important.

Teaching to the NCTM Standards

The *NCTM Handbook of Research on Mathematics Teaching and Learning* devotes much attention to the effectiveness of cooperative learning. Some of the benefits cited in the research include "students' involvement in curriculum tasks increased during cooperative, small-group work."[19] However, "the opportunity provided during small-group instruction varies for different students."[20] The application below supports this notion. Problems must be selected with great care. Problems that are too challenging will not engage the lower-achieving students, who typically have the greatest difficulties adapting to the cooperative group model. The research also found "low achievers may be discouraged from active participation in tasks because the more academically competent group members are concerned with getting tasks completed as quickly as possible."[21]

Classroom Applications

One example of an activity where cooperative learning may be beneficial is to have students investigate (or solve) a problem such as the following:

> A straight railroad track passes near two towns. One town is 5 miles from the track while the other is 8 miles from the track. The two towns, which are on the same side of the track, are 18 miles apart from each other. The citizens of the two towns want to build a railroad station somewhere along the track so that it is equally accessible from the two towns. Where should the station be placed?

Discuss the kinds of considerations that might be important in the decision. Explain and justify the geometric solution to the problem. Explain which other variables (not given) might be necessary to give the best location for the station.

Precautions and Possible Pitfalls

Perceptions don't always match reality. When planning to use cooperative learning, be selective about the types of problems or tasks given to the cooperative learning groups. Research on the effectiveness of cooperative learning in mathematics indicates it works more suitably with some math problems and skills than others. However, teachers' perceptions may not reflect this selectivity, and teachers may (unadvisedly) use cooperative learning across the board. Easy problems are unlikely to work effectively in cooperative learning groups because there is less need to collaborate to help each other.

Source

Peele, A., & McCoy, L. (1994). *Perception of group work in mathematics.* Paper presented at the American Educational Research Association Annual Meeting, New Orleans.

STRATEGY 13: Encourage students to work cooperatively with other students.

What the Research Says

When students cooperate with other students they often get more out of learning than they do when working on their own or even when working with the teacher. When students are isolated from each other and compete with each other, they are less involved in learning, their learning is not as deep, and they have fewer opportunities to improve their thinking. Students who work cooperatively achieve at higher levels, persist longer when working on difficult tasks, and are more motivated to learn because learning becomes fun and meaningful.

Teaching to the NCTM Standards

The NCTM Communication Standard stresses the importance of students becoming independent learners and students learning from each other. The concept of establishing a classroom community for learning is encouraged and valued as students become stakeholders in their own instruction program. The Communication Standard states: "The instruction program should enable all students to communicate their mathematical thinking coherently and clearly to peers, teachers, and others."[22] The *Handbook of Research on Mathematics Teaching and Learning* (NCTM) suggests that students are too passive and need to become more involved intellectually in classroom activities. As the application below suggests, breaking students into small groups engages more students in the learning process and can lead to a deeper understanding by a greater number of students.

Classroom Applications

Working cooperatively in a mathematics classroom can be done in many ways. Students can be given a challenging problem and then asked to solve it in small groups. While working cooperatively in a group setting, students must verbalize their thoughts and discoveries, which helps them to understand these ideas and use them as steps on the path to a solution. Teachers may present classical challenges such as filling in the cells of a 3×3 grid to form a magic square. Or, they may simply ask open-ended questions where mathematical solutions include judgmental aspects. For example, some real-life problems of this nature could be to determine the best location for a railway station, the minimum sum distance from town to station to town, and other conditions that can be included to describe a real situation. These factors present a problem that can be solved in a variety of ways by making judgments about the relative importance of variables. Such open-ended problem situations provide excellent opportunities for students to work cooperatively, probing and prodding each other in their quest for a solution.

Precautions and Possible Pitfalls

Not all group work is cooperative learning! Students must work together, help each other, and learn from each other in order for it to really be cooperative learning. Beware that one person doesn't dominate the group, which often tends to occur. Assigning group roles is one way of either preventing this from occurring or handling it when it does. Teachers need to constantly monitor groups as they are solving problems to make sure they stay on task and are working in productive ways.

Source

Johnson, D., & Johnson, R. (1975). *Learning together and alone: Cooperation, competition, and individualization.* Englewood Cliffs, NJ: Prentice Hall.

STRATEGY 14: Use group problem solving to stimulate students to apply mathematical thinking skills.

What the Research Says

Students interacting with other students when solving problems in a group stimulate basic (cognitive) and higher-level (metacognitive) mathematical thinking skills. A study was conducted with twenty-seven average-ability seventh graders in an urban school. A total of seventy-three problem-solving behaviors were examined for each student. One week before the study, students were put into heterogeneous groups so they could become familiar with how to work in groups and with the members of their groups. Heterogeneous groups were balanced for ability, gender, race, and ethnicity. During the study itself, no time limit was given for groups to solve the assigned problem. Groups were videotaped as they solved the problem. Working in groups to solve the problem, students engaged in the following types of mathematical thinking:

- Reading the problem (basic)
- Understanding the problem (higher level)
- Analyzing the problem (higher level)
- Planning an approach to solve the problem (higher level)
- Exploring a problem-solving approach to see whether or not it works (basic and higher level)
- Implementing the plan for solving the problem (basic and higher level)
- Verifying the final solution (basic and higher level)
- Listening to and watching other students during the problem-solving process

The greatest percentage of higher-level mathematical thinking occurred while students were exploring a solution; the second greatest occurred as students were trying to understand the problem. The highest percentage of basic mathematical thinking occurred during exploration; the second highest occurred while students were reading the problem.

Teaching to the NCTM Standards

The NCTM Professional Standards for Teaching and Learning state:

> Creating an environment that supports and encourages mathematical reasoning and fosters all students' competence with, and disposition toward, mathematics should be one of the teacher's central concerns. The nature of this learning environment is shaped by the kinds of mathematical tasks and discourse in which students engage.[23]
>
> Problem solving, reasoning, and communication are processes that should pervade all mathematics instruction and should be modeled by teachers. Students should be engaged in mathematical tasks and discourse that require problem solving, reasoning, and communication.[24]

Teaching problem solving as a group activity enables members of the class to analyze, compare, contrast, examine, and test solutions. "Students should be encouraged to explain their reasoning process for reaching a given conclusion or to justify why their particular approach to a problem is appropriate."[25] The application outlined below clearly addresses the standards and provides a stimulating format for meaningful problem solving.

Classroom Applications

Carefully select problems that will stimulate use of mathematical thinking and problem solving when preparing for group activities. Observe groups as they are working and intervene only as needed. After most groups seem to have finished, call the groups back together and have them describe their problem-solving processes and answers. Call attention to the mathematical thinking and problem-solving processes they were using, both when conferring with individual groups and after all groups have been called back together. Explicitly use the eight concepts described above when discussing problem solving to increase students' awareness of how they are thinking mathematically and how they are solving problems. Don't set rigid time limits for solving the problem. Let each group work at its own pace.

Precautions and Possible Pitfalls

Not all groups will behave in the same way. Some groups will not engage in all eight of the problem-solving behavior categories described above, and those that do may engage in them in varying degrees. In addition, not all groups will be successful in solving a given problem. If one group (or more) finishes before the others, make sure you have a follow-up task ready so students can extend their thinking rather than get bored and waste time waiting for the others to finish.

Source

Artzt, A., & Armour-Thomas, E. (1992). Development of a cognitive-metacognitive framework for protocol analysis of group problem solving in mathematics. *Cognition and Instruction, 9*(2), 137–175.

STRATEGY 15: Use the Jigsaw Technique of cooperative learning as an interesting and effective way for students to learn.

What the Research Says

Contrary to some beliefs about cooperative learning having only social benefits, research shows that the Jigsaw method helps students learn and apply academic content as well. An experimental study was conducted with seven classes of students in Grades 7 and 8. The 141 students were separated into four experimental classes and three control classes. The experimental classes were taught with the Jigsaw Technique, while the three control classes received regular instruction through lectures. The experiment lasted about four weeks, one double lesson per week. This study examined the social, personal, and academic benefits of Jigsaw and traditional instruction. Social and personal benefits observed to result from the Jigsaw method are growth of self-control, self-management, ambition, independence, and social interaction. Jigsaw was also found to reduce intimidation in the classroom, which inhibits learning and leads to introverted student behavior. The academic benefits of Jigsaw include improved reading abilities, systematic reproduction of knowledge, ability to make conclusions, and summarizing.

Students in the Jigsaw classrooms demonstrated improved knowledge as well as their ability to apply that knowledge when compared with students in traditional classes. Students were not afraid to ask questions or to scrutinize presented information when they were able to ask for and get an explanation of something from a peer.

Teaching to the NCTM Standards

The NCTM *Handbook of Research on Mathematics Teaching and Learning* discusses "the use of small cooperative groups of heterogeneous ability."[26] "The efficacy of small cooperative groups for increasing mathematics achievement in the general population seems well established."[27] The use of the Jigsaw techniques as described below engages all members of the class and clearly supports the communication standard as students are communicating their mathematical thinking and ideas to their peers.

Classroom Applications

The Jigsaw Technique operates in six steps:

1. Separate a new part of the curriculum into five major sections.
2. Split a class of twenty-five students into five groups of five students each. These groups are the so-called *base groups*. (The groups should be heterogeneous in terms of gender, cultural background, and achievement levels.)

3. Every member of the base group selects or is assigned one of the major sections. For example, one member might focus on the section on fractions, another might focus on the section on decimals, another focus on the section on percentages, and so forth. If the number of group members exceeds the number of sections, two students can focus on the same section.

4. The base groups temporarily divide up so each student can join a new group in order to become an "expert" in her or his topic. All the students focusing on fractions will be in one group, all the students focusing on decimals will be in another group, and so on. These students work together in temporary groups called *expert groups*. There they acquire the knowledge about their topic and discuss how to teach it to students in their base groups.

5. Students return to their base groups and serve as the expert for their topics. Everyone then takes a turn teaching what he or she learned about his or her topic to members of his or her base group.

6. A written test is given to the entire class.

In Steps 4 and 5, students get an opportunity to discuss and exchange knowledge. Step 6 gives the teacher an opportunity to check the quality of students' work and to see what and how much they learned from each other. One of the advantages of this method of cooperative learning is that in Jigsaw there is always active learning going on and students do not become bored while passively listening to reports from other groups, as sometimes happens with the Johnson and Johnson "Learning Together" method.

Jigsaw can be used to teach a series of unrelated skills such as factoring, reducing or simplifying algebraic fractions, as well as topics that could be used to tie seemingly unrelated topics together, such as verbal problems.

Precautions and Possible Pitfalls

While students teach members of their base groups in Step 5, teachers are frequently tempted to join in the discussions and advise students regarding the best way to teach the subject to their base group. This type of teacher intervention prevents the social and intellectual benefits of Jigsaw. Although a teacher has to monitor groupwork in order to intervene when there are substantial mistakes in understanding the academic content, the teacher should not interfere with how students decide to teach this content to their peers.

Sources

Aronson, E., Blaney, N., Stephan, C., Sikes, J., & Snapp, M. (1978). *The Jigsaw classroom*. Beverly Hills, CA: Sage.

Renate Eppler & Günter L. Huber. Wissenswert in Team: Empirische Untersuchung von Effeckten des Gruppen-Puzzles (Acquisition of knowledge in teams: An empirical study of effects of the Jigsaw-techniques. *Psychologie in Erziehung und Unterricht,* 37 Jg., (1990) S. 172–178.

2

Enhancing Teaching Techniques

STRATEGY 16: Find out about your students' motivation regarding mathematics, and use that knowledge to refine your instruction.

What the Research Says

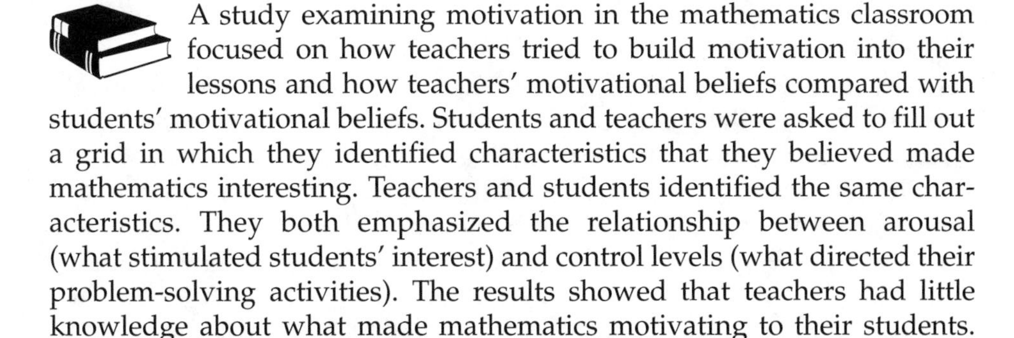

A study examining motivation in the mathematics classroom focused on how teachers tried to build motivation into their lessons and how teachers' motivational beliefs compared with students' motivational beliefs. Students and teachers were asked to fill out a grid in which they identified characteristics that they believed made mathematics interesting. Teachers and students identified the same characteristics. They both emphasized the relationship between arousal (what stimulated students' interest) and control levels (what directed their problem-solving activities). The results showed that teachers had little knowledge about what made mathematics motivating to their students. Findings from this study suggest that

- teachers need to learn what makes mathematics interesting to students,
- teachers need to pay attention to individual differences in student motivation,
- when teachers know about their students' motivational beliefs, they are more capable of refining their instruction so that students are interested in mathematics.

Teaching to the NCTM Standards

The NCTM Teaching Principle states: "Effective mathematics teaching requires understanding what students know and need to learn and then challenging and supporting them to learn it well."[1] Since teachers are living their lives in a different universe from students, it is highly recommended that teachers do everything to learn about their students' interests. When teachers know what interests students, they can plan lessons that employ that knowledge. In many instances psychological motivation can also be effective. Teachers can motivate students by imploring them to learn a little more so that they can complete a topic or chapter. Furthermore, teachers can motivate students by highlighting that mastery of mathematics is essential for success in other subjects and for success on important standardized tests for college admission.

The NCTM Connections Standard calls for students to "recognize and apply mathematics in contexts outside the classroom."[2] You may wish to regularly ask students to write about the relevance of each lesson to their daily lives. You will be delighted to learn of the "surprise" connections that students make with mathematics. As you prepare lessons in the future, you will have a rich assortment of "connections" that can motivate students.

Classroom Applications

The concept of intrinsic motivation means using interests already present in the learner to generate motivation or excitement in the subject matter. From this very definition, it is implicit that the teacher develops sensitivity for what interests students. This can vary from region to region and from student to student, and may vary with age, gender, and cultural background as well. There are, however, some relatively universal factors that many people harbor as interests; for example, the concept of completion. Students have the desire to feel that they have completed a task or topic and have relatively complete command of a concept. When teachers can craftily have students realize that their mastery is not yet complete, but with a little bit of further study it can be complete, they will have used a classic technique for motivating many students. A more complete treatment of this motivational technique can be found in Posamentier, Smith, and Stepelman (2006).

Precautions and Possible Pitfalls

Teachers need to cultivate the ability to determine what really motivates their specific students instead of assuming students are motivated by the same things that motivate the teachers themselves. To motivate students effectively, problems and topics must be appropriate in their content, structure, and level of difficulty.

Sources

Middleton, J. A. (1995). A study of intrinsic motivation in the mathematics classroom: A personal constructs approach. *Journal of Research in Mathematics Education, 26*(3), 254–279.

Posamentier, A. S., Smith, B. E., & Stepelman, J. (2006). *Teaching secondary school mathematics: Techniques and enrichment units* (7th ed.). Upper Saddle River, NJ: Prentice Hall.

Schiefele, U., & Csikszentmihalyi, M. (1995). Motivation and ability as factors in mathematics experience and achievement. *Journal of Research in Mathematics Education, 26*(2), 163–181.

STRATEGY 17: When trying to determine how to motivate students' interest in mathematics, teachers should differentiate between personal and situational interest and use both forms to increase students' motivation to learn mathematics. Teachers also need to both stimulate and maintain their students' interest.

What the Research Says

Teachers can draw on different types of interest that students have in mathematics. Personal interest is what students bring with them to the classroom or other environment; situational interest is something that is acquired by participating in an activity in the classroom or another situation. Whereas personal interest emphasizes the importance of working with individual differences in motivation, situational interest emphasizes the importance of the teacher's creating an appropriate setting to develop the students' interest in mathematics. Teachers should also differentiate between factors that stimulate student interest and those that maintain student interest. Computers, puzzles, and group work tend to stimulate interest in mathematics while meaningfulness and involvement tend to maintain student interest.

In a study of 350 high school students from three high schools, students were administered an interest survey with seven scales: personal interest, situational interest, meaningfulness, involvement, puzzles, computers, and group work. Students rated items on a six-point scale ranging from *strongly disagree* to *strongly agree*. Most students were white and were in thirteen college preparatory algebra or geometry classes. The results showed that situational interest in secondary school mathematics classrooms is complex, having five different components: meaningfulness, puzzles, computers, group work, and involvement. Increasing student involvement in mathematics appears to be especially beneficial for enhancing situational interest.

Teaching to the NCTM Standards

The NCTM Connections Standard states that students should "recognize and apply mathematics in contexts outside of mathematics."[3] Student learning is most effective when students are motivated. When teachers can incorporate personal interests of students in the mathematics that is being taught, students can better appreciate mathematics because it plays a role in their daily lives. When presenting topics that are not among students' interests, teachers must create or manufacture a "situational" motivating component. Each of the eight techniques listed below can effectively motivate students to learn a lesson.

Classroom Applications

There are many techniques for creating a situational motivation in the classroom. Sometimes the motivation lies in the material and other times it is dependent on the manner in which the activity is presented. Here are eight techniques for motivating a lesson:

1. Indicate a void in the students' knowledge.
2. Present a challenge.
3. Show sequential achievement.
4. Indicate the usefulness of a topic.
5. Use recreational mathematics.
6. Tell a pertinent story.
7. Get students actively involved in justifying mathematical curiosities.
8. Use teacher-made or commercially-prepared materials or devices.

For example, when motivating students about a topic (or lesson) on digit problems in the algebra class, or when the class is beginning to understand the workings of the decimal system (algebraically), you might have students try to explain the mathematical novelty:

> Why does the following arithmetic always result in the same number, 1089?
>
> Do the following:
>
> Choose any three-digit number (where the units and hundreds digit are not the same).

Subtract the number with the digits reversed.

To this difference, add the numbers with the digits reversed.

Your result should be 1089. Why?

Precautions and Possible Pitfalls

The primary precaution when doing a motivational activity is to make sure that it is appropriate for the intended students in both interest and level. In addition, ensure that it leads to the topic for which you are motivating your students rather than distracting them from it. If successful, try to modify other lessons to maximize student interest so that the usual classwork does not become boring compared to this highly interesting activity.

Source

Mitchell, M. (1993). Situational interest: Its multifaceted structure in the secondary school mathematics classroom. *Journal of Educational Psychology, 85*(3), 424–436.

STRATEGY 18: Treat students in ways that reflect the belief that you have high expectations for their performance.

What the Research Says

Although research has not supported the common belief that when teachers let students know that they have high expectations for their performance, a self-fulfilling prophecy tends to occur, research does show that teacher expectations can influence student performance in another way. When teachers have high expectations for their students' performance, they tend to treat students in different ways from when teachers have low expectations for student performance. When teachers believe their students can perform at higher levels, they tend to give students more encouragement or more time to answer a question than when they have low expectations of student performance. As a result of the increased time and encouragement, students tend to achieve at higher levels.

Teaching to the NCTM Standards

The NCTM Teaching Principle states that

> teachers establish and nurture an environment conducive to learning mathematics through the decisions they make, the conversations they orchestrate, and the physical setting they create. Teachers' actions are what encourage students to think, question, solve problems, and discuss their ideas, strategies, and solutions. The teacher is responsible for creating an intellectual environment where serious mathematical thinking is the norm. More than just a physical setting with desks, bulletin boards, and posters, the classroom environment communicates subtle messages about what is valued in learning and doing mathematics. Are students' discussion and collaboration encouraged? Are students expected to justify their thinking? If students are to learn to make conjectures, experiment with various approaches to solving problems, construct mathematical arguments and respond to others' arguments, then creating an environment that fosters these kinds of activities is essential.[4]

The research cited above does not link high expectations with high performance. However, creating a classroom environment that supports the goals of the NCTM Teaching and Learning Principles enable students to learn and have a flexible knowledge of mathematics. Thus, teachers who simply have high expectations without providing an environment that supports achievement will not enjoy the benefits of guiding a class to higher performance.

Classroom Applications

It is not uncommon that a student does not feel too comfortable with absolute value inequalities, especially at first exposure. First, there is the problem of understanding (or conceptualizing) the notion of absolute value, then there is the requirement of understanding how to express absolute value concepts in terms of inequalities. In short, it is not a terribly easy topic or concept for many students to master. To avoid students' becoming frustrated, teachers should show high expectations of students as a form of motivation. Consequently, this is a particular topic where teacher patience and encouragement is very important. Patience and encouragement convey to the student that the teacher believes she or he can succeed. Praising performance for this topic may have a particular value since the topic can be formidable for some students.

Precautions and Possible Pitfalls

Teachers should set realistic expectations, not expectations at a level so high that they are perceived as unattainable.

Source

Good, T., & Brophy, J. (1984). *Looking in classrooms* (3rd ed.). New York: Harper & Row.

STRATEGY 19: Praise mistakes!

What the Research Says

Mistakes are desirable! Most teachers consider mistakes as something forbidden. They immediately correct mistakes in the text, on the blackboard, on posters, in exercise books, and in every student's answer. Prohibiting mistakes produces anxiety about making mistakes, and accordingly, students are inhibited. This attitude toward mistakes easily can become a heavy burden, especially when there is time pressure.

In one study students and teachers ($N = 38$) were asked to solve the following problem within two minutes and to write down the answer.

> A customer buys a pocketknife for $6.00. He pays with a $10 bill. Because the owner of the shop does not have enough change, he goes to his neighbor and gets change for the bill. He gives $4.00 in change back to the customer. After the customer left the shop the neighbor came over and said, "The bill is counterfeit! I want my $10 back!" The shop owner quickly gave a real $10 bill to his neighbor. How much loss does the shop owner sustain if he bought the pocketknife at a price of $5.00 and if he does not count his loss of income?

Teachers' Answers					
Number of Teachers	5	7	12	9	5
Answer	$10.00	$15.00	$19.00	None	Correct

> The distribution of teachers' answers is equivalent to the students! Teachers gave the following reasons for making their mistakes: time pressure, feeling controlled, being in public, and lack of concentration.

One study (Bangert-Drowns, Kulik, Kulik, & Morgan, 1991) has shown that feedback is most effective when it is provided in a supportive manner, with an emphasis on guiding students to modifying their answers.

Teaching to the NCTM Standards

The NCTM Communication Standard states that students should "communicate their mathematical thinking coherently and clearly to peers, teachers, and others"[5] and "analyze and evaluate the mathematical thinking and strategies of others."[6] Teachers who encourage participation will have classrooms filled with students eager to contribute. In order to foster and maintain this healthy learning environment, teachers need to deal effectively with "wrong answers" in a thoughtful and meaningful way. The feedback a teacher provides to a student is a major factor in the student's learning process. Feedback should focus on how a student arrived at an answer. Teachers should not dwell on wrong answers, but instead highlight the "correct" thinking that the student employed in arriving at an answer. Almost all answers have redeeming qualities, and teachers should seek to find something good in every answer. Teachers should not draw additional attention to a student's incorrect response. Such actions are humiliating and do not foster an atmosphere where students are encouraged to volunteer, attempt alternative strategies, or take academic risks.

Classroom Applications

This kind of stress happens to almost every student in almost every lesson. Therefore, a teacher has to be careful when reacting to students' errors that result from stress. This works if the teacher reacts by giving small doses of help, so that the student can partially solve the problem and has some degree of success experience. In addition, you can use mistakes in creative ways, such as blueprints for student self-correction and as wrong answer choices for a multiple-choice test!

Creating anxiety in students about making mistakes depends on the teacher's reaction to mistakes. Teachers generally react on different levels. The first level involves informing the student that a mistake has been made. Subsequent levels usually involve giving substantial or formal help. The teacher's feedback can assume a variety of forms, from insulting to neutral to encouraging.

Examples of insulting the one who made mistakes:

- "Nonsense! Pay better attention!"
- "Sleepyhead! Your answer is rubbish!"
- "If you open your mouth, just rubbish comes out!"

Examples for neutral marking of mistakes, without evaluating the student who made them:

- "It is not right!"
- "That's wrong!"
- "There is some mistake!"

Examples of encouraging the one who made a mistake:

- "I am afraid that was not quite correct!"
- "Almost right! Try it again!"
- "Good idea, but, unfortunately not the right direction!"
- "Unfortunately wrong! If you continue thinking about it, you certainly will get the right answer!"

Creative Use of Mistakes

Mistakes can be used to uncover wrong ways of thinking. Therefore you should give problems that have several plausible solutions. Suggestions for solutions can be reasoned and discussed. Diagram 2.1 shows this schematically. Such discussions about how to solve the problem can occur in groups. Problems with several plausible solutions are quite rare. It is also possible to work with mistakes if students know the correct solution. Then they have to analyze their mistake. Diagrams 2.1 and 2.2 show this schematically.

Diagram 2.1

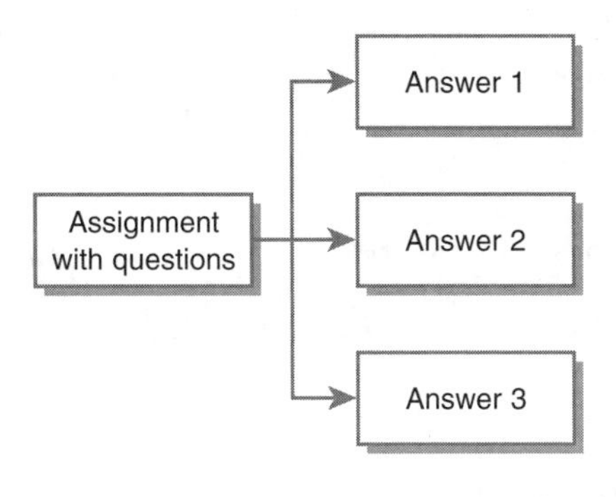

Why is Answer 1, 2, or 3 Correct?

Why is Answer 1, 2, or 3 Wrong?

Diagram 2.2

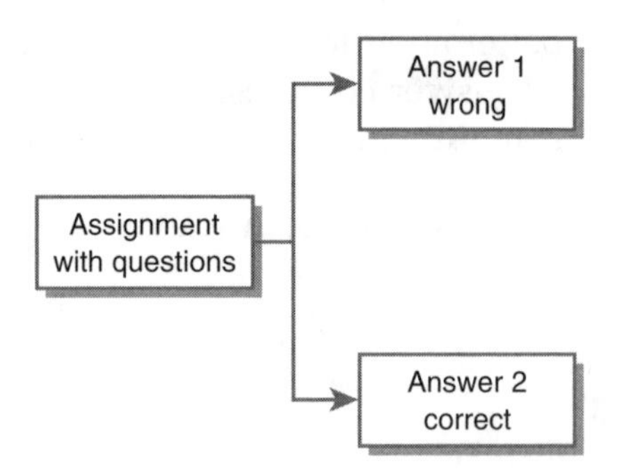

Why is Answer 1 wrong?
How is the wrong answer argued? What was the (wrong) way of thinking? What was mixed up?

Precautions and Possible Pitfalls

Do not wait for students to make mistakes to use errors creatively. It is recommended that you give students problems that are likely to cause errors so you can teach students to analyze them and use errors to improve their future performance. If you wait to use one of the many accidental mistakes that occur, students may interpret the detailed discussion that ensues as insulting or embarrassing the student who made the error, and the class might rebel against the wonderfully beneficial activity of error analysis.

Sources

Bangert-Drowns, R. L., Kulik, C. C., Kulik, J. A., & Morgan, M. (1991). The instructional effect of feedback in test-like events. *Review of Educational Research, 61*, 213–238.

Holger Morawietz. Fehler kreativ nutzen, Streß verringern, Unterricht öffnen (Creative use of mistakes, reducing stress, opening of lessons). *Pädagogik und Schulalltag*, 52 Jg., Heft 2 (1997) S. 232–245.

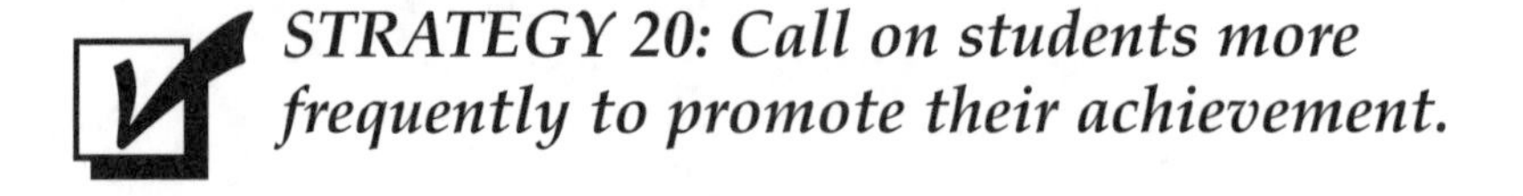

STRATEGY 20: Call on students more frequently to promote their achievement.

What the Research Says

A complex investigation of several interaction patterns in relationship to achievement in mathematics has shown, among other things, a strong connection between the frequency of being called on and the results of a geometry test. A total of ninety-two sixth-grade students (forty-four girls and forty-eight boys) took part in the experiment. The test was constructed as a combination of multiple-choice items and drawings. Students who were called on frequently got better results in both parts of the test than students who were not called on frequently.

Teaching to the NCTM Standards

The NCTM Professional Standards for Teaching and Learning ask that teachers

> Promote students' confidence, flexibility, perseverance, curiosity, and inventiveness in doing mathematics through the use of appropriate tasks and by engaging students in mathematical discourse. . . . Assessing the teacher's fostering of students' mathematical disposition should focus on whether the teacher facilitates students' flexibility, inventiveness, and perseverance in engaging mathematical tasks and on whether students demonstrate confidence in doing mathematics. Verbal cues that encourage students during instruction and supportive written comments on homework and test papers are obvious means of promoting a disposition to do mathematics. Teachers should be nonjudgmental when students give answers or present solutions to problems; teachers should help students to correct mistakes, but mistakes should be recognized as a natural part of the learning process.[7]

Maintaining a lively classroom that is punctuated by student questions and responses creates a positive learning environment. Students should be encouraged to ask and answer questions. Provide supportive feedback as you deal with answers that are both right and wrong. Involve all members of the class in inquiry and conjecture.

Classroom Applications

If you ask a question or assign a task and only a few students raise their hands, who should you call on?

a. The first student who raised his or her hand?
b. The student who raised his or her hand for the first time this day?
c. The student who is involved in the lesson most actively?
d. The student who guarantees a correct answer?
e. The student who does not raise his or her hand, but is known to be intelligent?
f. The student who is too reserved or anxious to raise his or her hand?
g. The student who seems to be inattentive?

The list could be continued almost indefinitely. This example illustrates that teachers' decisions about who they will call on is not so easy; however, the teacher has to decide within a very short time. It is impossible to always make the right decision. Whenever you call on one student, the rest may feel neglected. But you can pay attention and take control of the situation to make sure that you do not concentrate most of your attention on just a few students. Research has shown that teachers tend to pay more attention to students who are known as intelligent, such as those in (a), (c), (d), and (e) in the list above. Although students with learning weaknesses often give a wrong answer or no answer, it is necessary to invest the time in calling on them. Otherwise, their chance to achieve is reduced even further.

Precautions and Possible Pitfalls

Begin with calling on competent but reticent students. Then call on some of the weaker students. Sometimes when you call on students with learning weaknesses, they may give the wrong answer or no answer. In this situation the teacher frequently may call on another student who can give the right answer. If this happens to a weak student repeatedly, she or he is likely to lose self-confidence instead of gaining achievement or motivation. Instead of calling on another student for the correct answer all the time, at least some of the time you should guide the weaker student to the correct answer. This is likely to promote both achievement and motivation.

Source

Hans-Werner Heymann. *Lehr-Lern-Prozesse im Mathematikunterricht. Analysen im Bereich der Orientierungsstufe* (Processes of teaching and learning in math-classes). Klett-Cotta, Stuttgart 1978.

STRATEGY 21: Make sure to pause for at least four seconds after listening to a student's communication before responding.

What the Research Says

Extensive research for more than twenty years has compared extended pauses or "wait time" to normal pauses after a student's communication. Normal pauses average around one second. In a study of mathematics classes of twenty teachers covering the topic of probability reasoning, one group of teachers was taught to use extended pauses after listening to a student's communication, while the other group used their normal pauses. When teachers paused for at least four seconds after listening to a student's communication, before giving a response, middle school students had higher achievement in mathematics and were more likely to give thoughtful, detailed responses than students whose teachers' responses came immediately after the student spoke.

Teaching to the NCTM Standards

It is apparent that teachers who incorporate wait time into their questioning technique are rewarded with significant results. In addition to the benefit of obtaining extended responses, from a wider selection of students, teachers also gained valuable information about students' understanding of the concepts of the lesson. The NCTM *Research Companion to Principles and Standards for School Mathematics* states,

> One of the important ways that teachers assess student learning is through ongoing, everyday classroom use of questioning. As teachers and students engage in classroom discourse, teachers are obtaining important assessment information about where students are in their understanding. They can use this information to make decisions about the lesson, about the assignments, or about longer-range planning. Strong research evidence suggests that the nature of the discourse that occurs in a mathematics classroom can have a significant impact on students' learning.[8]

Classroom Applications

Instead of responding immediately to the student's communication, teachers should pause for at least four seconds after a student initiates a communication, such as an answer, question, or comment. This extended pause has the following benefits: (1) it allows students to think of a greater number of ideas; (2) it leads to students' thinking of more details about their ideas; (3) it gives more of an opportunity for other students to think about what the student communicating said, so another student can respond to the student's communication instead of the teacher; (4) it promotes higher levels of achievement in mathematics; and (5) it shows students, through teacher modeling, that it is appropriate to take time for reflection—to think about an answer before speaking.

Precautions and Possible Pitfalls

1. Teachers often respond quickly to students' questions and other communications. This rapid follow-up stifles higher-level thinking and the elaboration of ideas, and it cuts off students who were about to elaborate on their ideas.
2. This strategy sounds deceptively simple, but research shows it is hard for teachers to change their normal pausing habit and maintain a four-second pause before answering students.
3. The more complex the communication, the longer the pause should be.

Sources

Tobin, K. (1986). Effects of teacher wait time on discourse characteristics in mathematics and language arts classes. *American Educational Research Journal, 23*(2), 191–200.

Tobin, K. (1987). The role of wait time in higher cognitive level learning. *Review of Educational Research, 57*(1), 69–95.

STRATEGY 22: Use questions for different and versatile functions in the classroom.

What the Research Says

Research was conducted to investigate what questions teachers asked and why they asked them. Thirty-six high school teachers from five schools, representing all subject areas, participated in the study. They were asked to give examples of the questions they asked, to explain how they used them, and to tell to whom the questions were addressed. In general, the results showed that teachers ask questions to obtain information, to maintain control, and to test knowledge. Many specific functions of classroom questions were identified. These functions are listed below.

Teaching to the NCTM Standards

In the NCTM *Research Companion to Principles and Standards for School Mathematics,* the subject of using time effectively in the classroom is addressed to ensure that lessons are planned to allow for students to both "learn to communicate mathematically and to communicate to learn mathematics. This requirement calls for planning that involves capitalizing on what students do and directing their activities toward important mathematics issues."[9] Effective planning is not simply choosing the right number of problems to support the lesson, but, as the application below suggests, making important decisions about what to include in a given lesson so that meaningful learning takes place. "At the beginning of a discussion, the teacher might call on specific students selected in advance because he or she anticipates that a comparison of their solutions might lead to substantive mathematical conversation that advances the pedagogical agenda."[10] Allowing adequate time for student communication and involvement is key to planning a successful lesson.

The *Research Companion to Principles and Standards for School Mathematics* provides a great deal of evidence that supports the notion that classroom communication is an essential component of teaching and learning. The effective use of questions and responses serves many purposes in the mathematics classroom. The applications listed below highlight the many and varied uses of questions in the mathematics classroom. In addition to providing students a voice in the lesson, the skilled teacher can "revoice" the responses so that "the significant mathematical ideas could become the explicit focus of attention."[11] "Teachers revoice student contributions for a variety of reasons, including clarifying ideas, introducing new terms for familiar ideas, directing the discussion in a new and potentially productive direction, and helping students explicate their reasoning."[12]

> The teacher might reformulate a student's conjecture so that the grounds for it are expressed in a clear and coherent manner. In addition, teachers revoice students' explanations to bring them into relation with one another. . . . Alternatively, the teacher might revoice a sequence of responses so that they become a string of alternative ways of solving the same task.[13]

Classroom Applications

The following is a list of functions classroom questions can serve:

1. to stimulate curiosity or interest about a topic,
2. to concentrate attention on a specific concept or issue,
3. to actively involve students in learning,
4. to structure a task for maximal learning,

5. to identify specific problems interfering with students' learning,
6. to let a group of students know that their active involvement in the lesson was expected and that each person's contribution to the group was important,
7. to give students an opportunity to reflect on information and integrate it with what they already know,
8. to develop students' thinking skills,
9. to stimulate discussions in which students and the teacher develop and reflect on ideas, thereby enabling students to learn from others,
10. to show genuine interest in students' ideas and feelings.

Precautions and Possible Pitfalls

There are a variety of types of questions to avoid in order to make classroom questions maximally effective. However, there are certain techniques that also merit attention. For example, to maintain student interest in one another's responses, teachers should avoid repeating a student's response even if it was inaudible to the others. Have students who don't speak loudly repeat their response louder for others to hear. After a while, students will become so annoyed with the teacher's request to speak more loudly, they will begin to do it automatically and students will adopt the habit of listening to each other, realizing that the teacher will not repeat the response of a student. Another technique in asking questions properly is to avoid rephrasing a teacher's question for fear that its original rendition may have been unclear (unless students indicate the question is not clear). Surprisingly, there is usually more clarity to the original question, while unsolicited rephrasing might actually confuse those who understood the first version and now have trouble understanding the second version, thereby causing total confusion.

For additional pitfalls in questions see the chapter on "Classroom Questioning" in Posamentier and colleagues (2006).

Sources

Brown, G. A., & Edmondson, R. (1984). Asking questions. In E. C. Wragg (Ed.), *Classroom teaching skills.* New York: Nichols.

Posamentier, A. S., Smith, B. E., & Stepelman, J. (2006). *Teaching secondary school mathematics: Techniques and enrichment units* (7th ed.). Upper Saddle River, NJ: Prentice Hall.

STRATEGY 23: Teachers should be tactical in their use of questions.

What the Research Says

Thirty-six high school teachers from five schools, representing all subject areas, participated in a research study to investigate what questions teachers asked and why they asked them, giving examples of the questions they asked, an explanation of how they used them, and identification of to whom the questions were addressed. In addition, teachers were asked ten questions about their tactics for questioning. Five questions required the teacher to use a five-point scale to rate whether they agreed or disagreed with a tactic. Another five questions instructed teachers to estimate the frequency with which they used a questioning tactic. The "agree-disagree" results showed that teachers felt it was better to incorporate some or all of the items listed below. The "frequency of use" results showed that teachers

1. address questions to the whole class about half of the time
2. call on students by name about half of the time
3. praise students for correctly answering a question frequently, but not always
4. occasionally get students to ask each other questions
5. occasionally go all the way around the class, getting each student to answer a question

Teaching to the NCTM Standards

The NCTM Principles for School Mathematics in Teaching and Learning stress teacher understanding of what students know, and actively building new knowledge upon that foundation. Central to these principles is effective questioning techniques that can assess prior knowledge and foster links to higher-order conceptual understanding. Students should always be challenged to include "why" in their discussions and problem-solving strategies. Asking questions that require students to justify their approach to problem solving solidifies their understanding of the problem and provides provocative thought for other members of the class, who may be asked to concur with or refute the answer.

Classroom Applications

The following are just a few suggestions for questioning tactics:

1. After posing a question to the class, (silently) count to four or five to allow each member of the class to fully understand the question and formulate an answer.
2. Select a student to answer a question, rather than relying on volunteers.
3. Get another student to correct a student's incorrect answer instead of the teacher making the correction (students will begin to listen to each other's responses more carefully if they know that the teacher will not repeat a student's response).
4. Ask questions they already know the answer to only about half of the time.
5. Rephrase a question if a student doesn't understand it (after a reasonable waiting period) instead of redirecting the question to another student.
6. Orally discuss important questions rather than having students write the answers.
7. Audio- or videotape yourself while teaching to observe and evaluate your own questioning tactics; then develop a plan to improve your questioning tactics.

Far more space would be required here to treat this topic properly. We refer the interested reader to Posamentier and colleagues (2006) for a more in-depth treatment of the topic.

Precautions and Possible Pitfalls

Much of the above is dependent on teacher sensitivity. Remember, this is supposed to be a lesson-enhancing device. Do not get so hung up thinking about the teaching technique that it detracts from the lesson itself. Constant monitoring, not only of test results, but also an objective observation of student behavior, would be appropriate.

Sources

Brown, G. A., & Edmondson, R. (1984). Asking questions. In E. C. Wragg (Ed.), *Classroom teaching skills.* New York: Nichols.

Posamentier, A. S., Smith, B. E., & Stepelman J. (2006). *Teaching secondary school mathematics: Techniques and enrichment units* (7th ed.). Upper Saddle River, NJ: Prentice Hall.

STRATEGY 24: Make a lesson more stimulating and interesting by varying the types of questions you ask students.

What the Research Says

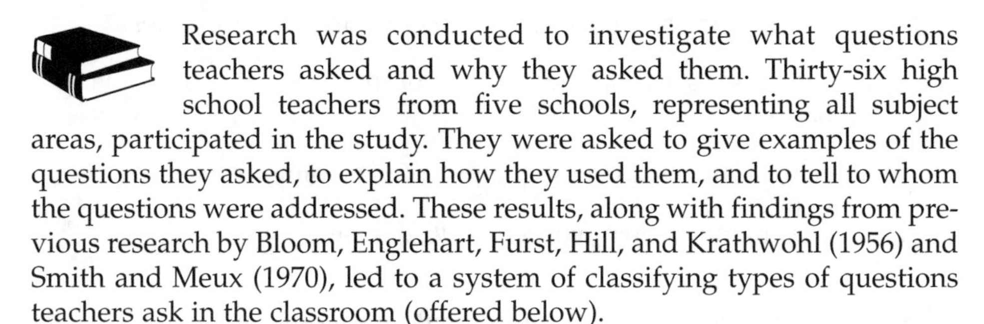

Research was conducted to investigate what questions teachers asked and why they asked them. Thirty-six high school teachers from five schools, representing all subject areas, participated in the study. They were asked to give examples of the questions they asked, to explain how they used them, and to tell to whom the questions were addressed. These results, along with findings from previous research by Bloom, Englehart, Furst, Hill, and Krathwohl (1956) and Smith and Meux (1970), led to a system of classifying types of questions teachers ask in the classroom (offered below).

Teaching to the NCTM Standards

The NCTM Reasoning and Proof Standard suggests that students should "select and use various types of reasoning and methods of proof."[14] Teachers, by varying their questioning techniques, can lead students to think more creatively and make valuable discoveries as the plot of the lesson unfolds through a thoughtful series of questions and responses. Questions should be constructed to stimulate different forms of thinking. Some (simple) questions merely test recalling of data or procedures while other questions aim to stimulate higher-order thinking and problem solving. Teachers' questions can shape student learning as the types of questions place emphasis on the process strands that are valued in the learning of mathematics.

Classroom Applications

There are many types of questions to use as well as many to avoid. Much of the non-drill topics in mathematics require understanding. When a topic that requires thought and deduction is being considered, it is wise to ask lots of questions. Each question should be succinct and structured in order to lead the students through a mathematical development or argument. One example is the sort of questioning that a teacher might use when guiding a student through a geometry proof. Here the questioning can take the form of factual questions or questions that have no definite answer but require a judgment to be made. Following is the list alluded to above.

Cognitive Questions

1. Recalling data, task procedures, values, or knowledge. This category includes naming, classifying, reading out loud, providing known definitions, and observing.

2. Making simple deductions, usually based on data that have been provided. This category includes comparing, giving simple descriptions and interpretations, and giving examples of principles.

3. Giving reasons, hypotheses, causes, or motives that were not taught in the lesson.

4. Solving problems, using sequences of reasoning.

5. Evaluating one's own work, a topic, or a set of values.

Speculative, Affective, and Management Questions

1. Making speculations, intuitive guesses, offering creative ideas or approaches and open-ended questions (that have more than one right answer and permit a wide range of responses).

2. Encouraging expressions of empathy and feelings.

3. Managing individuals, groups, or the entire class. This category includes checking that students understand a task, seeking compliance, controlling a situation, and directing students' attention.

Precautions and Possible Pitfalls

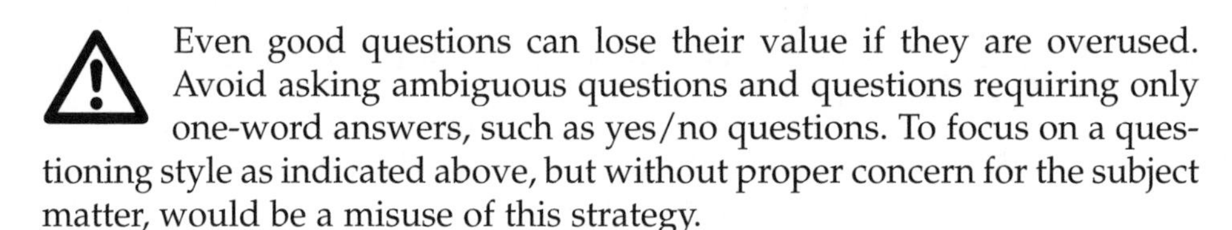

Even good questions can lose their value if they are overused. Avoid asking ambiguous questions and questions requiring only one-word answers, such as yes/no questions. To focus on a questioning style as indicated above, but without proper concern for the subject matter, would be a misuse of this strategy.

Sources

Bloom, B. S., Englehart, M. D., Furst, E. J., Hill, W. H., & Krathwohl, D. R. (1956). *Taxonomy of educational objectives: Handbook 1. Cognitive domain.* New York: David McKay.

Brown, G. A., & Edmondson, R. (1984). Asking questions. In E. C. Wragg (Ed.), *Classroom teaching skills.* New York: Nichols.

Smith, B., & Meux, M. (1970). *A study of the logic of teaching.* Chicago: University of Illinois Press.

STRATEGY 25: Use a variety of sequences to ask questions.

What the Research Says

As part of an investigation of questions teachers asked and why they asked them, thirty-six high school teachers from five schools, representing all subject areas, were asked to give examples of the questioning sequences they used and the context in which these sequences were used. Eight types of sequences emerged from the results:

1. *Extending:* A chain of questions on the same topic and of the same type
2. *Extending and lifting:* First the questions ask for the same types of examples, then there is a change to a different type of question, at a higher level
3. *Funneling:* Start with open questions and then get progressively narrower
4. *Sowing and reaping:* First a problem is posed and open questions are asked, questions become more specific, and finally the original problem is restated
5. *Step-by-step up:* Systematically move from recall to problem solving, evaluation, or open-ended questions
6. *Step-by-step down:* Systematically move from problem solving to direct recall
7. *Nose-dive:* Start with problem solving or evaluation questions and then move directly to direct recall
8. *Random walk:* No clear pattern of questioning content or type

Teaching to the NCTM Standards

In the same way that mathematical ideas "interconnect and build on one another,"[15] the NCTM Connections Standard suggests that an effective lesson can include the scaffolding of content through a carefully orchestrated series of questions. Each question should serve to provide a bridge to the desired aim of the lesson, make a connection to previously learned material, or challenge students to explore concepts beyond the scope of the lesson. Teachers can, with proper sequencing of questions, skillfully lead students through the process of justification of "how truth is established in mathematics."

Classroom Applications

Practice with the above forms of questioning sequences by considering one strategy at a time. Students can be guided nicely through mathematical proofs using an effective line of questioning; therefore, this may be just the right place to try these "skills." It should be noted that the questioning in doing proofs (most notably seen in the geometry course) can be channeled further to take into account the backwards strategy that should be used for all deductive proofs, no matter how simple. Students should be guided to begin at the end (the desired conclusion) and then work backwards, asking a variety of different types of questions until the given information is reached.

Precautions and Possible Pitfalls

Some of the questioning sequences listed above are better than others. Don't let students fall into the trap of thinking that all questions and questioning sequences are of equal value. Help them critically evaluate the value of each question and question sequence and consider situations in which each would and would not be desirable.

Source

Brown, G. A., & Edmondson, R. (1984). Asking questions. In E. C. Wragg (Ed.), *Classroom teaching skills.* New York: Nichols.

STRATEGY 26: Use a variety of strategies to encourage students to ask questions about difficult assignments.

What the Research Says

Several approaches have been identified that can help to overcome students' reluctance to ask questions:

1. Avoid giving students the impression that the reasons for difficulties are their own.

2. In cases where students have difficulties with problems, do not indicate that the problem was simple.

3. Give external reasons for students' difficulties.

Research has shown that people can handle their neediness better if they can attribute the reasons for their neediness to external causes. One study investigated asking questions as a kind of neediness. Participants were twenty-four girls and twenty-four boys, with a mean age of fourteen years.

Students were confronted with the following situation. They got an unformatted text that included typing mistakes. Students had to format the text according to a given pattern.

The results showed that:

1. Students showed the most willingness to ask questions when they could hold external circumstances responsible for their neediness.

2. Students' willingness to ask questions decreased when they had the impression that the person they asked blamed them for the difficulty. In this case, if the students asked a question, it would hurt their self-esteem.

3. Students avoided asking questions if the person they asked indicated the task was simple.

Teaching to the NCTM Standards

In the same manner that the NCTM Communication Standard requires students to think and speak mathematically, it is of equal or greater importance for teachers to create an atmosphere where students are encouraged to take risks as they pose questions and formulate responses to challenging problems in mathematics. Oftentimes individual students feel that they are alone in their inability to solve difficult problems when, in fact, it is a feeling shared by the entire class. Students are thus afraid to voice their uncertainty. *A Research Companion to Principles and Standards for School Mathematics* cautions teachers against silencing and marginalizing questions and responses from students that may be indicative of difficulty in understanding the concepts of a lesson. Instead, teachers should use the strategies outlined below to respond appropriately to student questions.

> In proactively supporting students' mathematical learning, the teacher necessarily has to treat students' contributions to the classroom discussion differentially. The decisions the teacher makes might be entirely justifiable from a mathematical point of view. However, in differentiating between students' contributions, the teacher implicitly communicates to students that certain opinions and ways of reasoning are particularly valued and others are less valued.[16]

Every question should be treated as a valued communication between students in the class and the teacher.

Classroom Applications

Teachers should explicitly and implicitly encourage students to ask questions. Asking questions is not easy for students in many cases. Sometimes even simple questions require both a minimum of knowledge/understanding and courage. Teachers should help students feel that there are no such things as silly questions, although teachers sometimes give silly answers! Asking questions is one of the most valuable skills a person can develop. Teachers can say that "silly" questions are often the very best questions. Teachers should:

- Make positive comments about students' questions

  Examples:

  "Good question!"

  "Instead of getting grades for good answers, you should get grades for good questions!"

  "Your questions show that you've thought about this a lot."

  "Very interesting question!"

- Encourage students to ask questions by emphasizing the difficulties of the task or of the working conditions. Or, give an understatement of your own abilities.

  Examples:

  "Some aspects of this problem are hidden. Consequently, you might have some difficulties."

  "We never even talked about some of the steps needed to solve this problem."

"I didn't even see this problem."

"Even today I have to struggle when asking questions in public."

- If students begin to attribute difficulties to their own lack of ability, try to direct their attention to the external difficulties.

  Examples:

  "Make sure you pay careful attention to the difficult parts of this problem."

  "This is a new type of problem. We haven't discussed it yet."

  "Do not expect your brain to work very quickly. It has been a long day."

- Do not express doubt about students' capabilities or skills.

  Negative Examples:

  "I already answered that question three times."

  "Listen carefully to what I say!"

  Positive Examples:

  "When students ask me a question a third time, that tells me that something has gone wrong with my explanation."

  "Okay! We covered a lot of facts—maybe too many."

  "Sometimes I explain things too quickly."

Precautions and Possible Pitfalls

⚠ Beware of possible backfire! When explaining an assignment's difficulties by external circumstances (very abstract, complex, obscure, or obtuse; or pressure for time or application of a very rarely used technique, etc.), you might encourage students; however, you can also confirm students' opinion that the assignment is too difficult anyway. In that case students would not be encouraged, but would feel justified in stopping work on the problem and questioning will cease!

Source

Urs Fuhrer. Fragehemmungen bei Schülerinnen und Schülern: eine attributierungstheoretische Erklärung. (Pupils' inhibition over asking questions: An attributional analysis). *Zeitschrift für Pädagogische Psychologie,* 8 Jg., (1994) S. 103–109.

STRATEGY 27: Use a Question-Asking Checklist and an Evaluation Notebook to help students become better learners.

What the Research Says

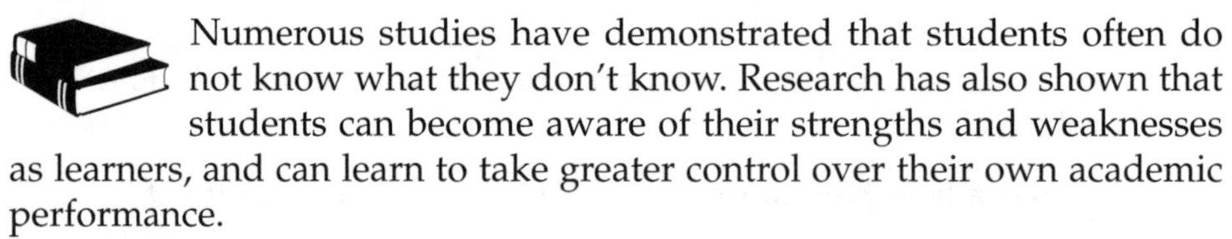

Numerous studies have demonstrated that students often do not know what they don't know. Research has also shown that students can become aware of their strengths and weaknesses as learners, and can learn to take greater control over their own academic performance.

One such study was conducted with sixty-four students from two ninth-grade science classes and one eleventh-grade biology class. There were four phases to this study:

- *Phase 1: Exploratory.* This phase lasted four weeks and involved getting to know the students and seeking their consent for cooperation and participation in the rest of the study.
- *Phase 2: Awareness.* This phase lasted five weeks for Grade 9 and three weeks for Grade 11. During this phase students began thinking about themselves as learners. They reflected on their attitudes, learning difficulties, and strategies for overcoming these difficulties.
- *Phase 3: Participation.* This phase lasted seven weeks for Grade 9 and six weeks for Grade 11. Students began using the Question-Asking Checklist and the Evaluation Notebook. The teachers gave students a considerable amount of help using these materials during this phase.
- *Phase 4: Responsibility-Control.* This phase lasted for seven weeks for Grade 9 and three weeks for Grade 11. During this phase the teachers' role virtually ceased and students used the materials on their own. Teachers monitored student behavior and attitudes and intervened only as needed.

Materials

- *Question-Asking Checklist.* There were ten different categories, each of which had its own icon and set of questions. The ten categories were topic, detail, task, approach, change in knowledge, increase understanding, progress, completion, satisfaction, and future use of knowledge. Questions for the approach category were "How will I approach the task?" "How hard will it be?" "How long will it take?" "Is there another way of doing it?" "Why am I doing the task?" "What will I get from it?" and "What will I make of the result?"
- *Evaluation Notebook.* Students evaluated their use of the questions from the Question-Asking Checklist for most of their science lessons during Phases 3 and 4 and recorded the results in this notebook. Data were collected from fifteen different sources, including notebooks, classroom observations, audio and video recordings of lessons, interviews, questionnaires, and teacher-made tests. The results showed that at the beginning neither the

ninth nor eleventh graders were clear about their learning difficulties, and they did not have strategies for overcoming them. During Phase 3 students became more aware of themselves as learners. During Phase 4 students improved in their ability to control their own learning.

Teaching to the NCTM Standards

The NCTM Learning Principle states that, "Learning with understanding can be further enhanced by classroom interactions, as students propose mathematical ideas and conjectures, *learn to evaluate their own thinking* and that of others, and develop mathematical reasoning skills."[17] While much emphasis has been placed on teacher questioning techniques, there should be equal emphasis on student self-questioning and self-evaluation. The application below provides a framework for student self-evaluation that can help students become less reliant on teacher assessment and provide their own formative assessment. This advances the goal of making them autonomous learners.

Classroom Applications

Such a checklist can also play an important role in the mathematics classroom when approaching a problem for solution. This list might include such questions as: "What does the problem call for?" "What am I to look for?" "What am I given?" "What do I know about the problem situation from prior experience?" "Where have I solved a similar or analogous problem?" "What is the relationship between the given data and that which is to be found?" "Is my answer reasonable?" and so on. The actual questions for the proposed problem must be consistent with the students' knowledge base and ability. By keeping a record and being forced to verbalize thoughts and actions, students come a long way toward getting involved in and reflecting on their own real problem-solving behaviors.

Precautions and Possible Pitfalls

Teachers must be careful not to impose their checklist on the students. They must have students formulate their own list so that the items are meaningful and useful to the students. Students should have a feeling of ownership of the list, and questions should be adapted to the specific problem.

Source

Baird, J. R., & White, R. T. (1984). *Improving learning through enhanced metacognition: A classroom study.* Paper presented at the annual meeting of the American Educational Research Association, New Orleans.

STRATEGY 28: Use school fundraising projects, such as students' selling candy, as the basis of mathematics lessons.

What the Research Says

Fundraising activities, such as students selling candy to raise money for a trip or special equipment, can be used as the basis of meaningful mathematics learning. One study examined how children develop and use mathematical knowledge through their experiences selling candy. This research showed that the hands-on experience of acting as a salesperson helped students to learn and understand important mathematical knowledge that they were later able to apply to working on school problems.

Teaching to the NCTM Standards

The NCTM *Handbook of Research on Mathematics Teaching and Learning* devotes a chapter to "Ethnomathematics and Everyday Cognition." In this treatise, ethnomathematics is defined to include mathematics that is involved in "everyday activities such as building houses, exchanging money, weighing products, and calculating proportions for a recipe."[18] Although the focus of the research showed how "non-mathematical" activities in the home required the use of mathematics and problem-solving skills, it can also be said that there are situations in school that occur outside of the mathematics classroom that support the use of mathematics. The application below suggests that students should participate in school activities that require the use of mathematics and support the classroom instruction, albeit in a nonconventional manner. These real-life activities may be more meaningful to students than the typical textbook exercises that are assigned for homework.

Classroom Applications

If there is a school store in your school, then it would be a good idea to make contact with the store manager and offer your class's services to do the accounting for the operation. This would then require getting all the purchasing and sales information about the store and then having the class decide how to manage the information. If there is no school store, then you might start one or get the principal's permission to undertake a project of fundraising. Funds so generated may be to improve the school, or to purchase important equipment for the school, such as band uniforms or computers. Students could also combine mathematics with social studies content by raising funds to be contributed to a local charity for homeless people. Once the project has been approved and the plan set, let students calculate the cost of doing business, the price of the items to be sold, and the anticipated profit based on specific sales results.

More sophisticated mathematics (or higher-order thinking skills) can be seen in students' estimating how much money they would generate if they offered different discount rates on the items they were selling. Students could compare the relative benefits of conducting different types of fundraising events, such as bake sales versus T-shirts. In addition, they could make projections for next year's fundraising target based on current data. Finally, they could compare the success of school fundraising strategies with those of nonschool fundraisers, such as local churches sponsoring benefit dinners or raffle ticket sales.

Precautions and Possible Pitfalls

Care must be taken to ensure that the activities undertaken support the mathematics program and that the "business" doesn't take on a life of its own, where the main (original) purpose to motivate and excite students about mathematics gets lost.

Source

Saxe, G. B. (1988). Selling candy and math learning. *Educational Researcher, 17*(6), 14–21.

STRATEGY 29: Don't give students feedback on their performance too early.

What the Research Says

Giving grades early stimulates students to participate actively in their lessons, but may undermine achievement in the long run. Previous research provided evidence that students learn because of anxiety over grades or because they get good grades with a minimum of efforts. Giving grades early is especially beneficial for students who require more time to understand things. They tend to be afraid of saying something wrong and of getting bad grades. Early grading is not viewed as judgmental of a student's knowledge; rather, it is viewed as informative.

This study investigated four ninth-grade classes on the effects of giving grades at an early stage of knowledge acquisition. To show the effects of early marking, four classes were separated into two groups. Both groups received computer-aided instruction and got a grade after every step. The first group did not get to know about grades, while the second group was informed about its grades. The achievements of the groups were compared on the basis of the grade after every step and on a final test. Students who knew their marks did slightly better on the interim tests. Their learning was enhanced by the grades. In contrast, on the final test, students who did not know their interim grades did noticeably better. They were not pushed by the pressure of marks. They used additional work to develop self-control. In this way, they dealt with the issue of their learning needs, understood the subject matter profoundly, and achieved at higher levels.

Teaching to the NCTM Standards

The NCTM Professional Standards for Teaching Mathematics address the importance of appropriate teacher discourse in the mathematics classroom. "Deciding when to provide information, when to clarify an issue, when to model, when to lead, and when to let a student struggle with a difficulty"[19] are important decisions that have a profound effect on the learning cycle. As the application below suggests, struggling students who are graded quickly may be frustrated and tune out. Students who are achieving well may feel that they no longer have to apply themselves toward mastery of the content.

Classroom Applications

Avoid giving grades at an early stage of learning. Students who are not interested in a particular topic or even the whole subject can easily get frustrated by early marks, and their motivation can sink even farther. On the other hand, early grades can promote rapid success. However, in some cases this leads to students' resting on their laurels. During the period when students are acquiring new knowledge, use grades sparingly.

Precautions and Possible Pitfalls

Do not stop all assessment during the early stage. First of all, students need assessment to evaluate or at least estimate their own achievement. In addition, you will always find some students who are motivated entirely by grades. Therefore, during the early learning phase, you should use oral or nonverbal assessment techniques.

Source

Hans Joachim Lechner & Ralf-Ingo Brehm, Meger Zbigniew. Zensierung und ihr Einfluß auf die Leistung der Schüler (Influence of marks on students' achievements). *Pädagogik und Schulalltag,* 51 Jg., Heft 3 (1996) S. 371–379.

STRATEGY 30: Use homework as a way of delving more deeply into important mathematical concepts and skills.

What the Research Says

Mathematics teachers in the United States put too much emphasis on breadth of content coverage and not enough emphasis on depth. This is one explanation for the finding that eighth-grade students in the United States rank below the international average in mathematics. This finding came from the Third International Mathematics and Science Study, which examined achievement results of 500,000 seventh- and eighth-grade students from forty-one countries. Researchers also studied textbooks and curriculum guides used throughout the world to teach mathematics and science. The study showed that students in the United States are introduced to more topics in mathematics and science than are students from other countries, but these topics generally are not covered in depth. Approximately 60% of mathematics teaching time per week is based on textbooks that were described as "a mile wide and an inch deep." This study suggests that teachers need to extend learning beyond the classroom to a greater extent than they do currently.

Teaching to the NCTM Standards

The NCTM Assessment Standards provide a clear rationale for assessment. "Teachers monitor students' progress to understand and document each student's growth in relation to mathematical goals and to provide students with relevant and useful feedback about their work and progress."[20] The NCTM suggests that students' progress is to be evidenced by a variety of sources, and homework is one means by which teachers can measure the mathematical growth and maturity of their students. Unlike typical classroom tests whose time constraints oftentimes limit the nature and quality of questions that are presented, homework assignments provide an outstanding opportunity for students to explore topics in depth, and at a pace that is comfortable for them. The typical mathematics lesson, due to its developmental nature, saves the most challenging (and interesting) applications for the end of the lesson. The teacher is oftentimes racing against time to complete this application and, as the period ends, many students do not fully comprehend it. As the application below suggests, a carefully planned homework assignment that culminates with problems that require a deeper understanding of mathematics is an advisable model to follow as students can explore these applications in a more relaxed fashion. The beginning of the next day's lesson can be spent clarifying any remaining issues.

Classroom Applications

Plan homework assignments as carefully as you plan your classroom lessons. Both the quality of the homework assignments and the amount of time students spend doing their homework are important for improving achievement in mathematics. An assignment that merely asks students to do, say, Exercises 2–20 (even numbers only) on a specific textbook page is one that does not show much thought. It is not an activity that usually enables students to deepen their knowledge of the concepts presented. Rather, it simply provides practice with a skill. Because classroom time is limited, use homework to give students more experience in understanding and applying important concepts and skills.

Precautions and Possible Pitfalls

Don't fall into the rut of making the same homework assignments year after year for class after class. Tailor your homework assignments to the needs of the specific students and classes. Take into consideration how much time students were actively engaged in classroom learning and how much time was spent on classroom management or other administrative tasks.

Sources

Checkley, K. (1997). International mathematics and science study calls for depth, not breadth. *Education Update, 39*(1), 1, 3, 8. Alexandria, VA: Association for Supervision and Curriculum Development.

Schmidt, W. (1997). *A splintered vision: An investigation of U.S. science and mathematics education*. Hingham, MA: Kluwer.

STRATEGY 31: When doing inquiry lessons, give students clearly written materials to guide the inquiry process.

What the Research Says

Research on the inquiry-based computer program "Geometric Supposers" was conducted in twenty-three high school geometry classes. This program, "Geometric Supposers," uses teacher-posed inquiry problems. The researchers specifically designed the materials so they would clearly communicate to the students what a particular problem is and what appropriate inquiry activities are expected. Evidence was collected from six sources: classroom observations, student interviews, students' work on the Supposers program, teacher interviews, teacher reflections, and minutes from monthly teachers' meetings. The results showed that (1) clearly written materials mean students will understand what work needs to be done, (2) they help students organize their work, (3) using charts and tables that tell students which measurements to make and giving step-by-step instructions was ineffective because it limited their inquiry. These findings led the researchers to conclude that how inquiry materials are written affects students' success in inquiry.

Teaching to the NCTM Standards

The NCTM *Handbook on Mathematics Teaching and Learning* devotes much attention to learning geometry at various grade levels using inquiry-based lessons. The use of dynamic software and computer environments shows results that are "intriguingly consistent. Difficulties and misconceptions that are easily masked by traditional approaches emerge in computer environments."[21] Thus, as the application below suggests, students must be given a precise roadmap upon which the inquiry can be based. Appropriately designed inquiry lessons can provide a high level of conceptual understanding of geometric ideas, and the students will achieve a sense of autonomy in their mathematical thinking.

Classroom Applications

Teachers should prepare clearly written materials for students, making sure to leave an opportunity for students to conduct an inquiry. Three strategies for writing clear materials are as follows:

1. State the goal of the problem at the top of the page.

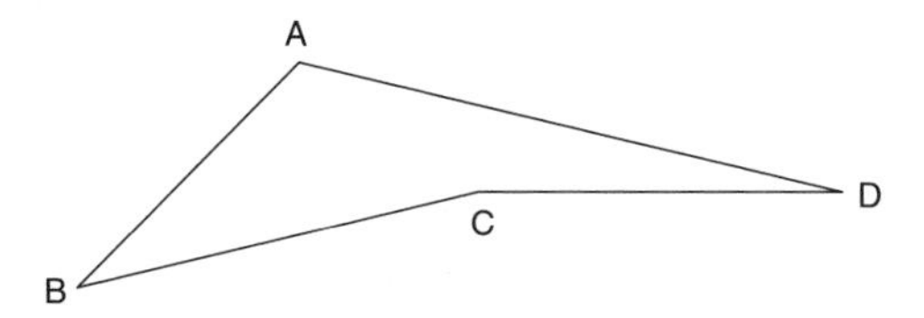

2. Provide explicit instructions on the processes to use when solving the problem so students know how to engage in inquiry; but don't provide so much structure that it cuts off student inquiry and it turns into students' collecting data or following directions without understanding what they are doing.
3. After students understand how diagrams can be used as models, use diagrams to simplify written instructions.

Sample Goal Statement

Task: To develop a procedure that enables you to reproduce or reconstruct Figure 1.

Sample Process Instructions

Procedure:

Make a drawing similar to this figure.

Collect data.

Describe below the procedures for reproducing this figure.

State your conjectures.

Precautions and Possible Pitfalls

Written material that is not clear prevents students from making successful inquiries. When creating inquiry problems for students, teachers should consider three issues before writing the problem statement and three issues while writing it. The three issues to consider before writing the problem statement are the kind of problem, its scope, and the students' background or ability. Three issues to consider while writing the problem statement are stating the goal of the problem, describing any constructions in the problem, and giving instructions for inquiry processes students are expected to use.

Source

Yerushalmy, M., Chazan, D., & Gordon, M. (1990). Mathematical problem posing: Implications for facilitating student inquiry in classrooms. *Instructional Science, 19,* 219–245.

3

Facilitating Student Learning

STRATEGY 32: Use inquiry-based learning in addition to problem-based learning.

What the Research Says

Tell me and I forget, show me and I remember, involve me and I understand.

—Chinese proverb

Numerous studies have demonstrated that students obtain a true understanding of concepts when they are involved in their discovery. Teachers in the Vermont Elementary Science Project observed the actions of students involved in hands-on science exploration. They found and enumerated the following characteristics of students who were involved in inquiry-based science learning:

1. Children viewed themselves as scientists in the process of learning.
2. Children accepted the "invitation to learn" and readily engaged in the exploration process as evidenced by increased curiosity and a willingness to try out their own ideas.

3. Children planned and carried out investigations. This included designing experiments, setting criteria for drawing conclusions, and collecting and interpreting data.
4. Children communicated in a variety of ways including sharing ideas; writing about their work; and drawing diagrams, graphs, and charts.
5. Children provided explanations and solutions and asked important questions that they themselves pondered.
6. Children asked important questions and were thoughtful observers.
7. Children critiqued their scientific method in an effort to modify it and assess its effectiveness.

Teaching to the NCTM Standards

In *Principles and Standards for School Mathematics,* the NCTM states that instructional programs should enable students to

> build mathematical knowledge through problem solving; solve problems that arise in mathematics and in other contexts; apply and adapt a variety of appropriate strategies to solve problems; and to monitor and reflect on the process of mathematical problem solving.[1]

This, coupled with the overarching theme of developing autonomous learners, supports the notion that we want students to be active participants in the learning process by doing mathematics, formulating conjectures, and entering into a discourse about the validity of a conclusion. Involving others in the mathematical discourse makes the entire class stakeholders in the development of concepts and ideas that we view as our goal as mathematics educators.

Classroom Applications

While learning in the mathematics classroom must include the acquisition of skills and concepts, we also value the role of inquiry in learning. The role of teacher shifts slightly in the inquiry-based lesson to that of a facilitator, charged with the responsibility to develop critical thinkers capable of extending the inquiry process beyond the classroom and developing students to prepare them for logical thinking for the remainder of their lives.

The advent of technology affords the mathematics instructor a wonderful opportunity to have students discover relationships in mathematics. Geometry lends itself perfectly to inquiry-based lessons. In particular, dynamic software programs for geometry allow students to discover relationships, make conjectures, and defend generalizations. The topic of currency within a triangle can be easily taught using the inquiry-based learning model with dynamic software. Ask students to draw the three medians of a triangle using Geometer's Sketchpad. They will soon "discover" that they intersect in a common point. Ask them to "move" the vertices of the triangle and then make a conjecture. Most of the class will hypothesize that the medians are concurrent at a point (the centroid). Challenge them to discover some other relationship of the centroid to the medians. Some of the more astute (or lucky) students will discover that the centroid divides the median into segments whose ratio is 2:1. Again, have them drag the vertices of the triangle to determine if this relationship holds true for all medians, in all triangles. They should also be gently guided to consider the areas of the six triangles formed by the three medians. They will discover that they are all equal in area.

This lesson requires some old-fashioned technology too. The teacher should have a stiff piece of cardboard, a few pairs of scissors, and a class set of compasses. Groups of students should be given cardboard triangles with different characteristics. They should be instructed to draw the three medians of the triangle, identify the centroid, and discover something else about the centroid. Students are amazed to learn that the centroid is the center of gravity for the triangle.

Precautions and Possible Pitfalls

⚠ Teachers should not limit inquiry-based learning to those lessons that can be taught with the aid of technology. This lesson could have been taught the old-fashioned way with each child constructing the three medians with compass and straightedge. In addition, the final discovery could have been made and verified by each student. Technology is not always available, and we can effectively employ inquiry-based learning in many situations where technology is not available.

Sources

Concept to classroom, Thirteen Ed Online, Educational Broadcasting Corporation, 2004. Available online at http://www.thirteen.org/edonline/concept2class/w6-resources.html

Inquiry based science: What does it look like? In Synergy learning [Special issue]. (1995, March-April). *Connect Magazine*, p. 13.

STRATEGY 33: To reduce math anxiety, focus on both the thoughts and the emotions of the students.

What the Research Says

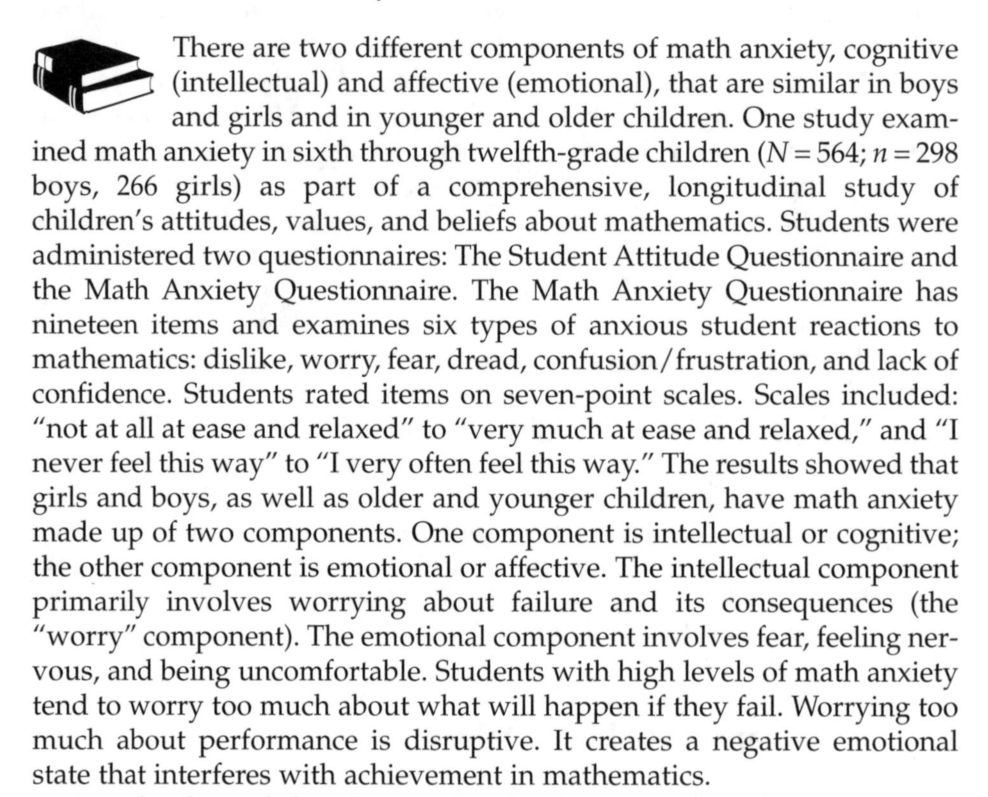

There are two different components of math anxiety, cognitive (intellectual) and affective (emotional), that are similar in boys and girls and in younger and older children. One study examined math anxiety in sixth through twelfth-grade children ($N = 564$; $n = 298$ boys, 266 girls) as part of a comprehensive, longitudinal study of children's attitudes, values, and beliefs about mathematics. Students were administered two questionnaires: The Student Attitude Questionnaire and the Math Anxiety Questionnaire. The Math Anxiety Questionnaire has nineteen items and examines six types of anxious student reactions to mathematics: dislike, worry, fear, dread, confusion/frustration, and lack of confidence. Students rated items on seven-point scales. Scales included: "not at all at ease and relaxed" to "very much at ease and relaxed," and "I never feel this way" to "I very often feel this way." The results showed that girls and boys, as well as older and younger children, have math anxiety made up of two components. One component is intellectual or cognitive; the other component is emotional or affective. The intellectual component primarily involves worrying about failure and its consequences (the "worry" component). The emotional component involves fear, feeling nervous, and being uncomfortable. Students with high levels of math anxiety tend to worry too much about what will happen if they fail. Worrying too much about performance is disruptive. It creates a negative emotional state that interferes with achievement in mathematics.

Results showed that the emotional component of math anxiety had a stronger and more negative relationship to children's perceptions of their ability and their performance, and to their actual math performance, than did the intellectual or worry component. The worry component had a stronger and more positive relationship to the importance children place on math, and their reported actual effort in math, than did the affective component. Girls reported stronger negative emotional reactions to math than boys did. Ninth graders reported they experienced the most worry about math and sixth graders reported the least amount of worry. There was relatively little change in math anxiety scores from junior through senior high school.

Teaching to the NCTM Standards

The NCTM Learning Principle calls for "Learning with understanding . . . learning the basics is important; however, students who memorize facts without understanding are not sure when or how to use what they know."[2] When students learn procedures without understanding their underlying mathematical principles, they lack the facility to apply these procedures broadly to become creative and competent problem solvers. Mathematical anxiety is a by-product of students' uncertainty about their abilities in mathematics. Teachers must look for signs of mathematical anxiety and reassure students that learning involves some frustration, but, with dedication, they are capable of mastering the content. Teachers are encouraged to have clearly defined performance indicators so that students do not "worry" about being held responsible for extraneous material. Students who have a clear understanding of what they are expected to know, and are treated in a compassionate, sensitive manner, display far less anxiety in the mathematics classroom.

Classroom Applications

1. Ask questions of students and listen to how they think about mathematics.
2. Analyze students' errors and identify recurring patterns. Help students convert failures to improvement plans so they will be more successful in the future.
3. Pay attention to body language as well as hidden verbal messages that reflect how students feel about mathematics in general, and how they feel about the particular problem they are working on. Show acceptance and don't be judgmental about their feelings. View students as capable of learning even when they doubt themselves.
4. Reeducate students about the learning process. Help them understand that learning is a long, slow, gradual process that involves trial and error, confusion, hard work, and failure as part of normal and natural learning. Forgetting things and making mistakes doesn't mean you're stupid.
5. Use everyday life applications so that mathematics isn't a scary and obscure mystery.

Precautions and Possible Pitfalls

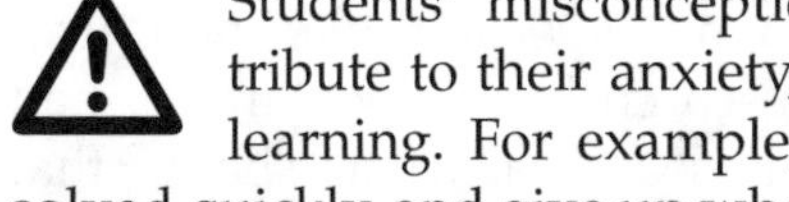

Students' misconceptions about the learning process may contribute to their anxiety, erode their confidence, and interfere with learning. For example, a student might expect a problem to be solved quickly, and give up when it isn't. A teacher's awareness of this syndrome can be helpful in dismantling it before it becomes deeply entrenched in the student as well as after it has already become deeply entrenched.

Sources

Gourgey, A. (1992). Tutoring developmental mathematics: Overcoming anxiety and fostering independent learning. *Journal of Developmental Education, 15*(3), 10–14.
Wigfield, A., & Meece, J. (1988). Math anxiety in elementary and secondary school students. *Journal of Educational Psychology, 80*(2), 210–216.

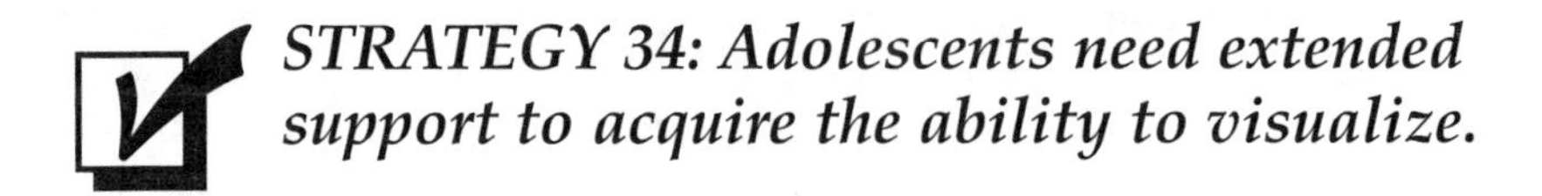

STRATEGY 34: Adolescents need extended support to acquire the ability to visualize.

What the Research Says

Boys and girls during infancy, childhood, and/or adolescence impart and acquire knowledge and emotions mainly in kinesthetic and auditory ways. This is very clear when you think of the way children or young people communicate. They speak in very plain, harsh terms and corroborate their words by slapping someone's back, giving a light shove, or making other gestures involving big motions.

In general, there are four methods of visualizing (see Figure 3.1).

Figure 3.1

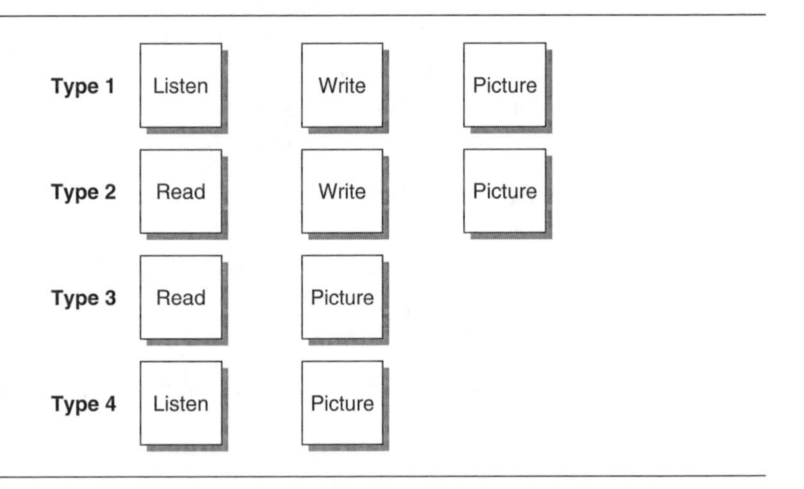

Previous research has shown that students who use all four types of visualization will learn more easily in school than students who remember only through auditory methods. Students who especially like the fourth type are very often regarded as gifted. Research has shown that urban students are primarily auditory learners. Research also has shown that the ability to visualize does not develop by itself. In addition to applying sketches and pictures during lessons, the nonverbal performance of the teacher is decisive for students' ability to visualize. The way a teacher speaks can improve students' visualizing—whether the students are already advanced in visualizing or not.

Teaching to the NCTM Standards

The NCTM's *Tools for Enhancing Discourse* encourages the use of concrete materials to support the auditory learning process. Included in this process are pictures, diagrams, tables, and graphs as well as technological tools like graphing calculators and dynamic software. Teachers must understand and value the use of visualization as a means to clarify information that is being presented. The visualization process provides students with another approach to communicating with others and intrinsically enables them to solidify important mathematical relationships. It also helps to deepen understanding by linking their experiences and informal knowledge to the mathematics they are learning in the classroom.

Precautions and Possible Pitfalls

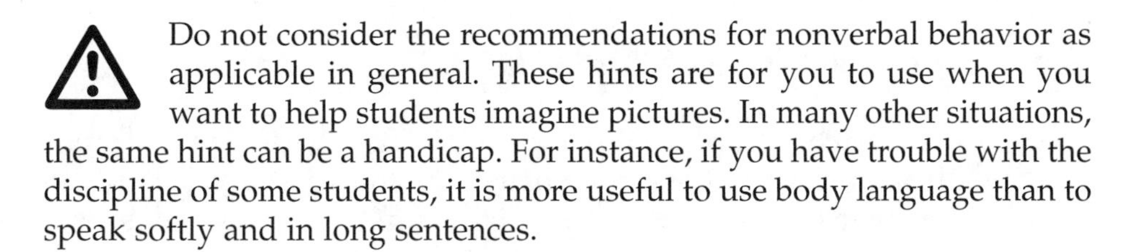

Do not consider the recommendations for nonverbal behavior as applicable in general. These hints are for you to use when you want to help students imagine pictures. In many other situations, the same hint can be a handicap. For instance, if you have trouble with the discipline of some students, it is more useful to use body language than to speak softly and in long sentences.

Source

Grinder, M. (1992). *NLP für Lehrer. Ein praxisorientiertes Arbeitsbuch* (NLP for teachers. A praxis orientated workbook). Freiburg im Breisgau, Germany: Verlag für Angewandte Kinesiologie GmbH.

STRATEGY 35: Use graphic representations or illustrations to enhance students' memory while they are listening to you. Abstract representations such as flow charts are more effective than colorful pictures.

What the Research Says

There are several types of supplementary materials that can help students remember what they learn while listening to a teacher. Some types of materials are better than others. One study compared the effect on students' memory of graphic representations (flow charts with keywords) versus colorful pictures. The study was conducted with twenty-three girls and thirty-three boys, ages eleven to thirteen years. The entire sample was separated into four groups. In each group students listened to a tape recording of textual material. The four types of material were:

Type	*Description*
1	No graphic representation and no picture
2	Graphic representation (flow chart with keywords from the text)
3	Colorful picture
4	Graphic representation and picture

After hearing the tape, students completed a questionnaire to check the effects of the different types on students' memory. The results showed that students remembered more when they were given the graphic representations.

Teaching to the NCTM Standards

A Research Companion to the NCTM Standards devotes much attention to graphical representation in an expository piece titled "Learning to Graph and Graphing to Learn." The focus of this piece centers upon using the graphical representations as a means to engage students in the "broader communication process." This process should support and complement the auditory process and must be carefully chosen not to interfere with it. The article concludes with the statement:

> Graphing is no longer a topic consisting of a few skills and procedures to be taught once and for all. As a means of communication and of generating understanding, graphing must repeatedly be encountered by students as they move across the grades from one area of school mathematics to another.[3]

Classroom Applications

Pictures can inhibit memory of knowledge given auditorily. The complexity of a picture takes too much visual attention and stresses the capacity of intellectual processing. When students read a text, then pictures can support the verbal information, because during reading a person can focus her or his attention either on text or on picture and can move ahead at her or his own pace. However, pictures can interfere with listening. They can interrupt the logical flow of the words. Pictures and words go together only when the auditory information explains the picture.

There are many types of graphic representations mathematics teachers can use. They include

- flow charts
- rough structural sketches
- continuums
- matrices
- Venn diagrams
- tree diagrams
- concept maps
- problem-solution charts

Graphic representations are characterized by: being quickly understood, providing a structure for integrating new information, and being schematized sketches instead of colorful pictures.

Figure 3.2

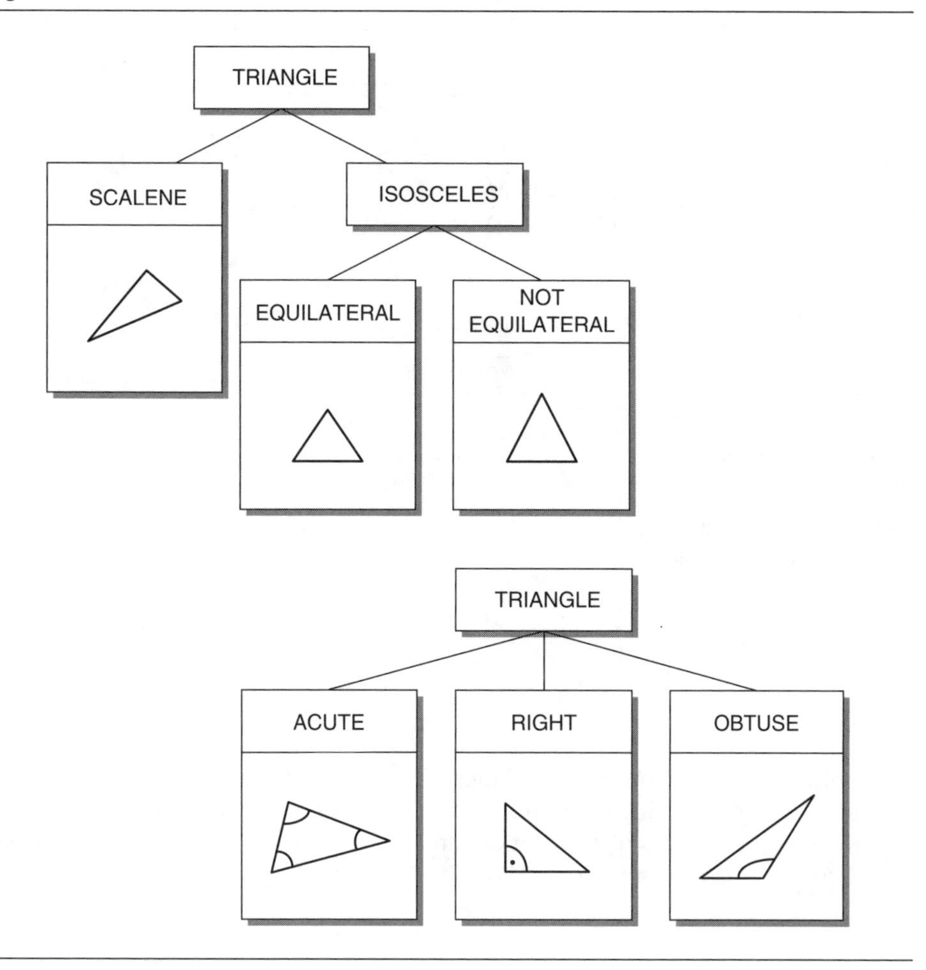

Precautions and Possible Pitfalls

The illustration in Figure 3.3 demonstrates that not every graphic representation aids learning and memory. The representation in Figure 3.3 draws too much attention away from what students will be listening to because it is too complex. It combines several characteristic classifications into one representation. That is why it is recommended to construct a graphic representation as simple as possible if students are to see it while they are listening. A complex representation such as the one in Figure 3.3 may be useful for students when they are not listening to a lesson.

Figure 3.3

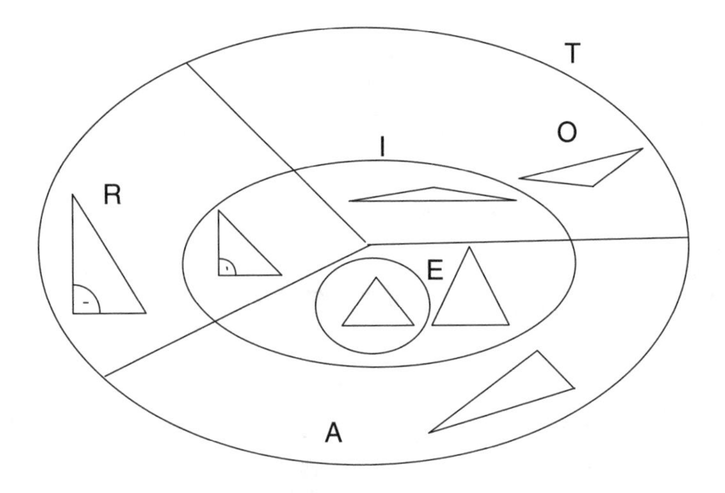

T: set of all triangles
A: set of all acute triangles
O: set of all obtuse triangles
R: set of all right triangles
I: set of all isosceles triangles
E: set of all equilateral triangles

Source

M. Imhof, B. Echtrernach, S. Huber, S. Knorr. (1996). Hören und Sehen: Behaltensrelevante Effekte von Illustrationen beim Zuhören (Ear and eye: Retention of effects of illustrations in listening tasks). *Unterrichtswissenschaft Zeitschrift für Lernforschung*, 24(4), 229–342.

STRATEGY 36: Teach students to ask themselves questions about the problems/tasks they are working on.

What the Research Says

When students ask themselves questions about the work they are doing and problems they are solving, thinking and learning are enhanced. Learning tends to occur somewhat differently in the various subject areas. Learning and problem-solving strategies used in writing are not exactly the same as those used in mathematics. However, in all subjects, students need to self-regulate their performance in similar ways. Research examining more than 100 videotapes of high school and college students working on unfamiliar problems showed that generally students are not aware of or do not self-regulate their problem solving the way an expert does. Students usually made quick decisions about how to approach a task and persist in that direction, whether right or wrong. If students are wrong in their initial idea about how to proceed, failure is guaranteed unless they look back, reconsider, and try another approach. Research has demonstrated that these types of self-regulation skills (metacognitive skills) can be taught to students through explicit instruction. Self-questioning is one effective strategy for self-regulating mathematical problem solving.

Teaching to the NCTM Standards

The NCTM Problem Solving Standard calls for students to "monitor and reflect on the process of mathematical problem solving."[4] The above-referenced research supports this learning standard because it is important for students to consider alternate strategies while solving a problem if the chosen strategy is unsuccessful. Students who consider a few problem-solving strategies before beginning are more likely to adopt an alternative strategy should their initial approach not succeed. In addition to the individual applications discussed below, *A Research Companion to Principles and Standards for School Mathematics* cites the work of Lampert (1990), who found it important to "organize whole-class discussions around students' answers to a problem and to call their answers 'conjectures' until the class as a whole discussed the legitimacy of various solution strategies."[5] An extension of this would be to have students critique their own work by modeling the whole-class discussions surrounding the validity of student conjectures.

Classroom Applications

Teach students to ask themselves questions before, during, and after they solve problems or work on other tasks. Their questions should be formulated to focus specifically on the problem or task at hand. Examples include, "What technique did I use to solve a similar problem in the past?" "How do I find the derivative?" "What is the problem asking for?" "What information am I given?" Students should also ask themselves general questions designed to self-regulate their performance, such as, "Is there anything I don't understand?" "Am I headed in the right direction?" "Is there any information given in the problem that is not immediately obvious?" "Have I made any careless mistakes?" In mathematics, the reasonableness of the answer obtained is often not considered. Students believe that mathematics problems are contrived, therefore, the solutions do not really apply to the real world. This self-regulating experience should be extended to this aspect of self-checking by considering not only the procedure used, but also the answer arrived at.

Precautions and Possible Pitfalls

Questions generated by the students themselves are more effective than questions provided to them by the teacher. Although student questions are often unpolished and may even sound inaccurate, students understand them and could resent having them sharpened by the teacher. Allow students grammatical and content freedom in their private self-questioning activity, or else you may spoil the genuineness of the experience with largely irrelevant, extraneous factors, however well intentioned they may be.

Sources

Lampert, M. (1990). When the problem is not the question and the solution is not the answer: Mathematical knowing and teaching. *American Educational Research Journal, 27,* 29–64.

Schoenfeld, A. (1989). Teaching mathematical thinking and problem solving. In L. B. Resnick & L. E. Klopfer (Eds.), *Toward the thinking curriculum: Current cognitive research.* Alexandria, VA: Association for Supervision and Curriculum Development.

STRATEGY 37: *Teachers can help students learn to ask better questions.*

What the Research Says

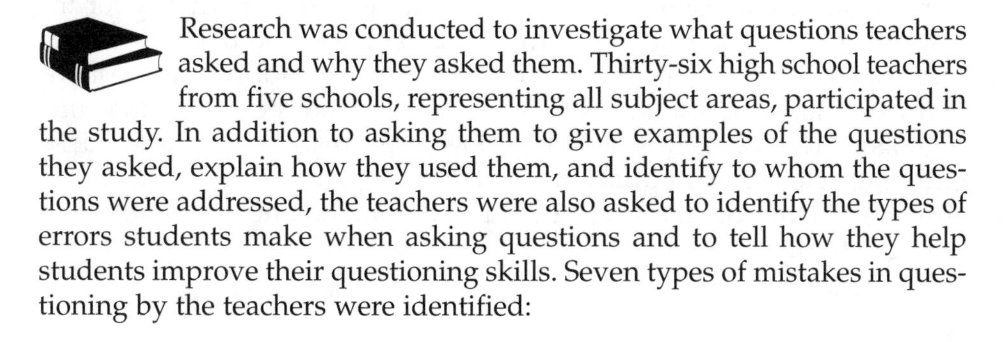

Research was conducted to investigate what questions teachers asked and why they asked them. Thirty-six high school teachers from five schools, representing all subject areas, participated in the study. In addition to asking them to give examples of the questions they asked, explain how they used them, and identify to whom the questions were addressed, the teachers were also asked to identify the types of errors students make when asking questions and to tell how they help students improve their questioning skills. Seven types of mistakes in questioning by the teachers were identified:

1. *Delivery:* Questions that were unclear; too fast, too slow, too quiet, or too loud; and not having eye contact with students while questioning
2. *Structure:* Questions with unclear demands, such as using unclear vocabulary, and making questions too long or too complex
3. *Target:* Questions directed inappropriately to an individual, a group, or the whole class
4. *Background:* Questions that were not put in the proper context of the lesson or problems with questioning sequences
5. *Handling answers:* Not allowing enough time for students to answer questions or accepting only answers that were expected
6. *Discipline and management:* Avoid having students all call out answers at the same time and making sure all students can hear both the questions and the answers
7. *Level:* Avoid asking questions that are too hard or too easy for the specific students targeted

Teachers made several suggestions for helping students ask better questions. They recommended providing students with models of good questions for students to observe, discussing effective questioning strategies, giving students opportunities to practice questioning, and providing students with feedback on their questioning.

Teaching to the NCTM Standards

The NCTM Communication Standard calls for students to "organize and consolidate their mathematical thinking through communication."[6] In the same fashion that effective questioning techniques employed by teachers can enhance instruction, appropriately phrased student questions can provide a meaningful opportunity for the teacher to both assess student progress and extend the boundaries and focus of the lesson. Students should be encouraged to formulate questions that reflect an understanding of the lesson's focus or pose questions that highlight areas of confusion. In posing questions, students should be encouraged to use appropriate mathematical language. This "participation structure" reinforces the goal of students' thinking and communicating mathematically.

Classroom Applications

Making students aware of what a proper question is may, in the long run, make them better learners and give them the ability to pose better questions. Teachers should demonstrate and discuss the characteristics of effective and ineffective questions. It may prove fruitful to have students critique each other's questions, either in pairs or in a group. Alternatively, give a list of student-generated questions to students and have them critique them as a homework assignment. There is a fair amount of information available about proper questioning techniques. Students need to know what makes effective classroom questioning as well as what could make the questions counterproductive. A good source on this topic would be Posamentier, Smith, and Stepelman (2006).

Precautions and Possible Pitfalls

Given peer pressures, students critiquing each others' questions may be a bit of a challenge for the teacher. Avoid embarrassing students for asking ineffective questions by calling attention to them in a whole-class setting. Teachers who attempt the above suggestion for implementation must be aware that this is meant to be an enhancement for the instructional program and not a deterrent. If one technique doesn't seem to work with a particular class, it ought to be replaced with a more effective approach.

Sources

Brown, G. A., & Edmondson, R. (1984). Asking questions. In E. C. Wragg (Ed.), *Classroom teaching skills.* New York: Nichols.

Posamentier, A. S., Smith, B. S., & Stepelman, J. (2006). *Teaching secondary school mathematics: Techniques and enrichment units* (7th ed.). Upper Saddle River, NJ: Prentice Hall.

STRATEGY 38: Give students the kind of feedback that will most help them improve their future performance.

What the Research Says

Teachers often give students less than useful information about their performance. Studies have shown that students benefit more from learning about when they are wrong than when they are right. In addition, for students to improve their future performance, they need to know why something is wrong. Research shows that teachers often fail to provide students with this kind of information about their performance. When students understand why something is wrong, they are more likely to learn appropriate strategies to eliminate their errors.

Teaching to the NCTM Standards

The NCTM Professional Standards for Teaching Mathematics endorse thoughtful discourse between the teacher and the student. When giving feedback to individual students or the entire class, teachers must be careful to organize the feedback in a manner that can best support the learning objectives. The application below shows how the teacher can differentiate the types of errors that students make so that effective feedback can be provided. Conceptual errors require feedback and support that are quite different from that required for careless calculation errors. Teachers should be careful not to overwhelm students with feedback to ensure that it is absorbed. Feedback should be prioritized to highlight those areas that the teacher feels are most crucial to the comprehension of the subject matter. Providing informative feedback, along with appropriate instructional support, will enable students to meet your learning objectives.

Classroom Applications

Mathematics instruction, in general, has lots of opportunities to give a right answer or a wrong answer. A "gray area" hardly exists. For a teacher to merely indicate the right answer or to indicate that a student's response is wrong does little to aim the student in the right direction. Teachers should analyze incorrect responses to see if the errors are in reasoning, incorrect interpretations, faulty work with algorithms, or the like. Though such an analysis can often be time-consuming, it is extremely worthwhile, for it is the discovery of the error (resulting from an error analysis) that can be the key to helping a student sort out his or her mathematics difficulties.

There are several types of errors that occur in the normal mathematics classroom. First, there are errors that are common to a large portion of the class. These can be attributable to a misunderstanding in class or to some prior learning (common to most of the class) that causes students to similarly react incorrectly to a specific situation. When the teacher notices this sort of thing, a general remark and clarification to the entire class would be appropriate. The misconception may be one of not understanding a concept, such as the average of rates not being treated as the arithmetic mean (because it is, in fact, the harmonic mean). Or it could involve the incorrect use of the quadratic formula by a number of students who carelessly do not draw their fraction bar long enough to include the numerator. Often, errors are simply due to careless mistakes. Teachers should help students develop and implement specific correction plans to prevent these errors from recurring in the future. We direct the reader to a pertinent discussion of student errors, "The Logic of Error" by Dr. Ethan Akin (1996).

Precautions and Possible Pitfalls

It is possible that several errors may be revealed by an error analysis of a student's work. To point out too many faults at one time could confound the student and consequently have a counterproductive effect. The teacher should arrange the discovered errors in order of importance and discuss them successively with the student, one by one, going on to the next only after successful completion of the earlier one. Teachers should follow up to see if students have successfully followed their error correction plans and have rectified previous errors, especially recurring errors.

Sources

Akin, E. (1996). The logic of error. In A. S. Posamentier & W. Schulz (Eds.), *The art of problem solving: A resource for the mathematics teacher.* Thousand Oaks, CA: Corwin.

Bangert-Drowns, R. L., Kulik, C. C., Kulik, J. A., & Morgan, M. (1991). The instructional effect of feedback in test-like events. *Review of Educational Research, 61,* 213–238.

STRATEGY 39: Help students understand their own thought processes and guide them in learning to think like mathematicians.

What the Research Says

Many students who have difficulty with mathematics tend to look at a problem, then quickly decide how to solve it. They proceed with their approach, whether or not it leads in the right direction. This impulsive tendency contrasts with the manner in which good mathematics students and mathematicians approach problems. They try to understand the problem, then carefully plan how to solve it, monitor the success of the approach in progress, and abandon it to find another if it does not lead them in the right direction. One study examined more than 100 videotapes of high school and college students working on unfamiliar problems. The results showed that more than 60% of the students used the impulsive approach described above. The research showed that when students impulsively decide on an approach that turns out to be wrong, fai ure is inevitable unless the students assess early enough whether the approach is working, and if it is not, change to a more effective solution strategy.

Teaching to the NCTM Standards

The NCTM Standards for Mathematics Teaching recognize the challenges that teachers face in getting students to understand their own thought processes. This is particularly more difficult with older students "who have become accustomed to a different set of standards for school thinking and talking." Students must "try to convince themselves and one another of the validity of particular representations, solutions, conjectures, and answers."[7] Creating a classroom atmosphere where students are encouraged to "use a variety of tools to reason, make connections, solve problems, and communicate"[8] should be the goal of every mathematics teacher. Students should be challenged by their peers to provide evidence that their conjectures have validity and that mathematical ideas are employed appropriately to set up and solve problems.

Classroom Applications

To help students become more aware of and take more control over their own thinking, teachers can coach or guide their thinking process so it becomes more like that of a mathematician. Such guidance or coaching can occur by asking them:

1. How would one solution approach apply to another problem?
2. Exactly what are they doing?
3. Why are they using this particular approach to solve this problem?
4. How do they know whether the approach is leading in the right direction?
5. Is there any other way of solving this problem?
6. How will they know if the answer they get is right?
7. What will they do with the answer once they get it?

Precautions and Possible Pitfalls

Teachers should take individual differences into account when applying the ideas listed above. Since students come to the classroom with very different learning habits and needs, this list may have to be modified and/or supplemented to take student differences into account.

Source

Schoenfeld, A. H. (1989). Teaching mathematical thinking and problem solving. In L. B. Resnick & L. Klopfer (Eds.), *Toward the thinking curriculum: Current cognitive research* (Yearbook of the Association for Supervision and Curriculum Development). Alexandria, VA: Association for Supervision and Curriculum Development.

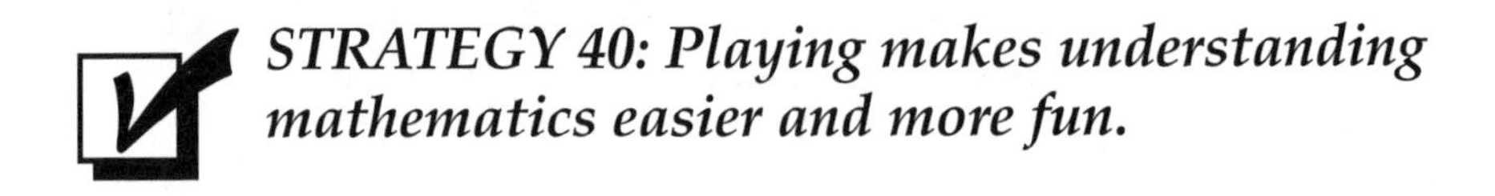

STRATEGY 40: Playing makes understanding mathematics easier and more fun.

What the Research Says

Research repeatedly has demonstrated the learning effects and motivating outcomes of lessons involving play. One study with third- to sixth-grade students examined the impact of play on achievement in mathematics. The results showed that in addition to students learning more, their active participation and motivation increased during the following class period.

Teaching to the NCTM Standards

The NCTM Problem Solving Standard calls for students to "apply and adapt a variety of appropriate strategies to solve problems."[9] Games are a very effective way to get students to engage in problem solving. Most games are based primarily in mathematics. Students who recognize the underlying mathematics and begin to strategize using mathematical representations are more likely to become proficient in both the game and the underlying mathematics. The game becomes a mental manipulative of the mathematics while the mathematical problem solving supports winning strategies of the game. Linking games to mathematics is simply the merging of informal knowledge with the mathematics of the day's lesson. It is a time-tested way to excite students about learning mathematics while highlighting the relevancy of the mathematics that they are learning.

Classroom Applications

The activity described below, which is like a game show, was used in the experiment described above. You may adapt it for your students and mathematical content.

1. Preparation for the Quiz

- Create a transparency with a basic diagram with thirty-six squares (see Figure 3.4).
- Six questions for luck and six questions for risk are distributed at random on a hidden copy of the basic diagram (see Figure 3.5). Make sure you have enough of the luck and risk questions. Students like them, and they motivate students to play.

Figure 3.4

The Big Prize

	1	2	3	4	5	6
A						
B						
C						
D						
E						
F						

Figure 3.5

The Big Prize

	1	2	3	4	5	6
A		*Luck*			*Risk*	
B				*Risk*		*Luck*
C			*Risk*	*Luck*		
D	*Luck*	*Luck*				
E			*Risk*		*Risk*	
F	*Luck*					*Risk*

2. Preliminary Round

- Every student has to answer three questions.
- Each student is allowed to personally select the mathematical topic (e.g., oral multiplication).
- Every correct answer earns two points for the student.
- Six points is the maximum per student.

3. Main Round

- Every student starts the main round with the number of points he or she has accumulated in the preliminary round.
- As soon as the moderator asks a question, students can raise their hands to answer.
- The student who raises his or her hand first is allowed to answer the question.
- The student who earned the previous points has to select the square for the next student.
- Each Luck square gives five points for the correct answer.
- Each Risk square allows the student to as much as double the number of points he or she has accumulated. The student decides how many points to risk. If the student correctly answers the question, the points that are risked will be added. If the student gives an incorrect answer, the number of risked points will be subtracted from his or her score.
- The remaining squares provide one point added for each correct answer and one point subtracted for each wrong answer.

4. Course of the Quiz (This can be given out to students)

- Class and teacher elect people for the following roles:
  - moderator (this could be the teacher)
  - assistant (covers the answered squares)

  - three counters (scorekeepers)
  - referee (decides on the order of students, who raised up their hands first)
  - six game players
- Arrange the chairs in a horseshoe. Players can sit at the front side and the "audience" at the other sides.
- The counters announce the score at the end of the preliminary round.
- The player who has the highest score starts the main round. In case of a tied score between several students, the winner is decided by the results of a play-off question.
- The transparency with the basic diagram (Figure 3.4) lies on an overhead projector. The moderator keeps the transparency that appears in Figure 3.5.
- The player who starts the main round chooses one square (e.g., square D4).
- The assistant covers this square with a stone or the like.
- The moderator asks the question.
- According to the order of raised hands, students give their answers until the correct answer is given. The referee observes the correct order and tries to see if anyone is cheating.
- If a square with Luck or Risk is selected, only the student who has chosen it is allowed to answer the question. Only if he or she gives a wrong answer does the question become open to the other players.
- The quiz is finished when all squares are gone.
- The student with the highest score wins.

Besides written tests, a teacher should look for fun and efficient ways to check knowledge, abilities, and skills.

Precautions and Possible Pitfalls

⚠ Playing games during lessons requires constant and strict organization, otherwise you will produce more problems than motivation. Give very precise directions for playing the game and for how you expect students to behave. In case students start to misbehave, stop the game and try it again during one of the next lessons.

Source

Otmar Borst. Der große Preis. Ein Spiel für alle Klassenstufen und Lerninhalte (The big prize. A quiz for all ages and all subject matter). *Mathematica Didactica,* 17 Jg., Heft 1 (1994) S. 106–118.

STRATEGY 41: Select and carefully structure homework assignments so that they require the development of mathematical thinking and reasoning. Anticipate changes that might occur while students are working at home.

What the Research Says

The mathematical tasks that teachers use have a major impact on the kinds of mathematical thinking that students do. Research has shown, however, that the tasks set up by the teacher may change, while students are working on them, due to complex factors in the environment. Such changes have been demonstrated to occur in the classroom and similar changes could occur at home. A sample of 144 mathematical tasks was analyzed in terms of features of the tasks and of the cognitive demands required by students engaged in these tasks. Features of the tasks analyzed included communication requirements (e.g., produce mathematical explanations or justifications), number of solution strategies, and number and kind of representations. Cognitive demands analyzed included memorizing, using procedures with and without connections to concepts, and doing mathematics. Researchers observed teachers announcing the tasks to students (setup) and observed students working on them (implementation). Observations occurred in three teachers' classrooms at four project sites over a three-year period, spanning sixth grade through eighth grade. Videotapes were made of the lessons to supplement observations. The results documented changes from setup to implementation in the cognitive demands, in the number of solution strategies, in the number and kind of representations, and in the communication requirements.

Teaching to the NCTM Standards

The NCTM Curriculum Standards embrace the notion that the curriculum should include numerous and varied experiences that reinforce and extend logical reasoning skills so that all students can

- make and test conjectures
- formulate counterexamples
- follow logical arguments
- judge the validity of arguments
- construct simple valid arguments.[10]

The homework assignment provides ample opportunities for students to practice achieving all of the above goals as they strive to attain mathematical power and become independent learners. As the application below clearly indicates, the homework assignment must be appropriately formulated so that the major concepts of the day's lesson are reinforced in the assignment. In addition, the homework assignment should attempt to extend the goals of the lesson and challenge students' mathematical reasoning abilities. Teachers must anticipate the challenges that students will encounter in completing the assignment and prepare them to overcome them. Students should be encouraged to explore topics beyond the scope of the syllabus, including those from the history of mathematics.

Classroom Applications

A good habit to nurture is to preview the homework assignment before the students work on it. In this way, students know exactly what is expected of them and where to place their emphasis at home. For example, if the homework assignment is to practice solving quadratic equations, then the teacher might elicit from students the various types of equations they may face, as well as the types of solution methods expected of them. The teacher might also ask students what type of format they would use, to make sure their thinking is headed in the right direction. Whenever possible, in order to promote the development of mathematical thinking and reasoning, teachers should elicit ideas and information from students instead of telling it to the students. If students aren't forthcoming with the appropriate information, then it is appropriate for the teacher to provide it and/or clarify ideas presented by students. It's a good idea to vary homework assignments nightly so students do not get into a rut where they do their assignments by rote instead of thinking about what they are doing and why. One possible assignment to stimulate mathematical thinking and reasoning could be to have students select a topic within an area of mathematics and research it for a presentation to the class. Students may also select a topic from the history of mathematics, such as how calculations were made in different periods of time, the significance of a particular mathematician's work, or the etymology of mathematical symbols or expressions. Such assignments should dovetail with regular classroom work.

Precautions and Possible Pitfalls

Don't preview the homework assignment to the extent that it might take some of the "suspense" or excitement out of the assignment. However, it is important that the teacher prepare the class for the assignment.

Source

Stein, M. K., Grover, B. W., & Henningsen, M. (1996). Building student capacity for mathematical thinking and reasoning: An analysis of mathematical tasks used in reform classrooms. *American Educational Research Journal, 33*(2), 455–488.

STRATEGY 42: Use homework assignments as opportunities for students to get practice and feedback on applying their mathematical knowledge and skills.

What the Research Says

Giving students numerous and well-designed homework problems to solve can help students improve their performance in mathematics by giving them opportunities for practice and feedback. Research indicates that many hours of practice are needed for students to be successful in mathematics. In a review of the literature of cognitive aspects of problem solving, the following benefits of practice with feedback were identified:

- Recollection of facts and concepts
- Recognition of patterns
- Recollection of strategies or procedures that can operate on patterns
- Use of strategies or procedures automatically
- Facilitation of work with problems presented in unfamiliar forms
- Analysis and identification of errors
- Development of problem-solving skills
- Development of important cognitive skills that cannot be explicitly taught
- Improvement of performance in creative problem solving, including solving problems that require "insight" or productive thinking rather than reproductive thinking
- Improvement of problem-solving speed

Teaching to the NCTM Standards

The NCTM Assessment Standards clearly outline a framework and purpose for monitoring students' progress. Because the goal of mathematics instruction is to create autonomous learners with increased mathematical power, classroom practice and assessment models must shift to accommodate these changing goals. Homework assignments must be used to challenge students and monitor students' attainment of mathematical power in addition to supporting basic skills. Additionally, homework assignments should be constructed to be used as one means of formative assessment; feedback from homework can be used to communicate progress to the student and it can inform the teacher of future instructional needs by highlighting skills and concepts that are in need of review, remediation, or reassessment. The application presented below shows how an experienced teacher might design a homework assignment that challenges all of the members of a class, and, at the same time, serves other purposes as well. The homework assignment given below also puts mathematics in context and supports the notion that topics in mathematics are connected. It also highlights how a homework assignment can and should prepare students for the challenges that await them in future lessons.

Classroom Applications

Homework assignments must be carefully designed. The exercises should be selected to provide a balance of problems solvable by all students, problems solvable by most students, and a few problems that may be given as a challenge for the better or more motivated students. Teachers should assign homework exercises that reflect a mixture of topics taught, thereby putting the various topics into the larger context of the course. Teachers might wish to spiral back over previously taught material so students get a sense of "context." They may also wish to foreshadow some topics by giving students a guided series of questions that lead directly to the next topic. For example, the teacher planning to introduce the proof of the Pythagorean Theorem on one day, may wish to have students identify the three mean proportional relationships involving the segments of the hypotenuse of a right triangle, when the altitude is drawn on the hypotenuse. The resulting proportions lead directly to one of the common proofs of the Pythagorean Theorem. You may wish to consult E. S. Loomis's (1968) classic book, *The Pythagorean Proposition,* for other proofs of this famous theorem, some of which could also be used to foreshadow, via a homework exercise, the proof of this theorem.

Precautions and Possible Pitfalls

Avoid the common teaching practice in mathematics of having students focus on the course topic by topic. Caution should be used for any homework assignment so as not to assign exercises that are merely repetitions of each other, but with different numbers used. This is not only extremely boring, but serves no purpose in the learning process. Often students do these by rote methods where *no* thinking is involved.

Sources

Frederiksen, N. (1984). Implications of cognitive theory for instruction in problem solving. *Review of Educational Research, 54*(3), 363–407.

Loomis, E. S. (1968). *The Pythagorean proposition.* Reston, VA: National Council of Teachers of Mathematics.

STRATEGY 43: Assign homework and other projects requiring students to write about connections between mathematics and other subjects.

What the Research Says

The Office of Educational Research and Improvement at the U.S. Department of Education funded the development of a digest to assist teachers in the difficult task of making connections between mathematics and other subjects. Research indicates that making such connections is especially difficult at the high school level, where students have different teachers for different subjects and there is a strong emphasis on distinct content areas. The National Council of Teachers of Mathematics recommends making such connections, but teachers often do not have the knowledge or resources needed to implement this recommendation. Research indicates that when students connect their mathematical knowledge and skills with other subjects, mathematics is seen as more interesting and more useful than when students see mathematics as a separate subject.

Teaching to the NCTM Standards

The NCTM Standards for Curriculum and Evaluation state that the "mathematics curriculum should include investigation of the connections and interplay among various mathematical topics and their applications so that students can use and value the connections between mathematics and other disciplines."[11] As the application below suggests, connections can easily be made to real-life applications. Sports and business sections of the newspaper provide real-life connections to mathematics. Asking students to discover other connections should be encouraged; students should find less obvious connections to mathematics, like the interpretation and manipulation of statistics to support differing positions and ideas.

Classroom Applications

Teachers often feel that they do not have enough time in class to help students connect what they are learning about mathematics with other subject areas, especially when they want students to read and write about these connections. Homework and other out-of-class assignments, such as reading newspapers and magazines, can be excellent ways of addressing this limitation. A resourceful teacher will be able to show students how almost every page of a newspaper contains applications of mathematics. Naturally, the sports pages and the business section are obvious illustrations of mathematics at work in subjects of gymnastics and history or social studies. These connections, although well known to most students, are not always obvious. For example, a "batting average" is actually not an average in the way mathematicians understand the term; rather it is a percentage or decimal. The most exciting discoveries of mathematics in use in the newspaper are those where the mathematics is embedded in the article, not necessarily those illustrations that give some quantifiers and the like. Asking students to regularly search for good newspaper applications of mathematics and their review in class are excellent ways of connecting mathematics across the curriculum. Students can also research and write about the relationships between mathematics and related topics in other subjects such as elections, taxes, and the stock market in history or social studies; mixtures of solutions and soil composition in science; fractals in art; as well as scales, chord structures, and chord progressions in music.

Precautions and Possible Pitfalls

⚠ One should be careful of using, as connectors to other fields, the trivial, or simple, illustrations of numbers being used to quantify a situation. Students may like to use these as their examples, but since they offer relatively little to connect mathematics to other fields, it would be wise to caution the class not to use these as their examples.

Sources

McIntosh, M. (1991). No time for writing in your class? *Mathematics Teacher, 84*(6), 423–433.

Reed, M. K. (1995). *Making mathematical connections in high school.* ERIC Digest. ERIC Clearinghouse for Science, Mathematics, and Environmental Education, Columbus, Ohio. (ERIC Document Reproduction Service No. ED380310 95).

Vatter, T. (1994). Civic mathematics: A real-life general mathematics course. *Mathematics Teacher, 87*(6), 396–401.

Wood, K. D. (1992). Fostering collaborative reading and writing experiences in mathematics. *Journal of Reading, 36*(2), 96–102.

STRATEGY 44: Consider whether a student's learning weakness might involve a deficiency in auditory perception.

What the Research Says

Sometimes learning difficulties are due to physical abnormalities rather than simple basic skill deficiencies. In order to discover possible causes of learning problems, research was conducted to investigate whether students' problems could be explained by disorders in auditory perception. The Mottier Test was administered to 104 students who were identified as having learning problems and had been treated by psychologists because of their learning problems.

The design of the test was as follows: The students had to listen to and repeat a total of thirty nonsense words (words without a real meaning). Words on the list became progressively longer and more complex. The results of the experiment revealed that all the students, whether they had repeated a grade or not, had problems in auditory perception.

Teaching to the NCTM Standards

The NCTM Communication Standard requires that teachers and students "communicate their mathematical thinking coherently and clearly to peers, teachers, and others."[12] When properly framed questions are posed and a particular student consistently displays difficulty in comprehending and responding to questions, teachers should consider other (nonmathematical explanations) for this shortcoming. One classroom practice that can help teachers diagnose both auditory and visual perception deficiencies is to rotate students' seating arrangements regularly and then carefully observe changes in student behavior. As students rotate to the rear of the classroom, both auditory and visual disorders will become more obvious and allow the teacher to consult with other teachers and guidance support personnel to make appropriate referrals for a professional diagnosis.

Classroom Applications

In addition to other possible causes, sometimes learning difficulties can signal deeper problems in auditory perception. Students who cannot understand what they hear will have problems comprehending information and directions given by the teacher. They will also be unable to follow the comments of other students. Hence, teachers should have some background knowledge in auditory perception problems in order to recognize the symptoms and assist in the remediation process. To determine whether students might have difficulties in auditory perception that are affecting their learning, teachers can use a test like the Mottier Test.

Precautions and Possible Pitfalls

It is not necessary to specifically use the Mottier Test in order to diagnose auditory perception problems. Teachers can construct their own tests, quite simply, with the same results. Although a teacher-constructed test will not be recognized as a valid test, it can provide teachers with a signal as to whether follow-up on auditory perception is worthwhile. Consideration should be given to the possibility of other causes, such as a temporary deficiency in students' concentration or difficulty in spelling. Administering the test two or three times is recommended in order to check whether the results reflected a serious problem or a chance occurrence. To determine whether problems in auditory perception or in spelling produced the learning problems, contact a language teacher for his or her experiences with the student concerned.

Source

Holger Wagner. Auditive Merkfähigkeit bei Schülern: Eine Studie zum Mottier-Test (Auditory memory in school-age children: A study about the Mottier test). *Psychologie in Erziehung und Unterricht*, 37 Jg., (1990), S. 33–37.

STRATEGY 45: Complex exercises that give students freedom tend to fit the way older students learn.

What the Research Says

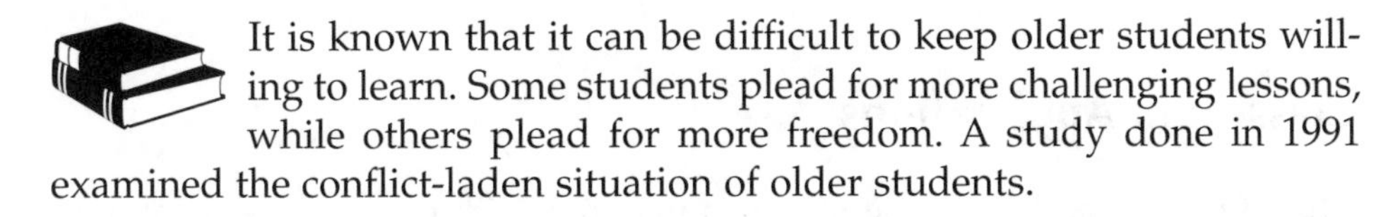

It is known that it can be difficult to keep older students willing to learn. Some students plead for more challenging lessons, while others plead for more freedom. A study done in 1991 examined the conflict-laden situation of older students.

- Senior students see school more and more as a place to meet people the same age; existing schools, however, do not meet their social needs.
- Students do see school as primarily a place of intensive and useful learning. However, teachers stress only learning in school and emphasize learning when trying to motivate students.
- Senior students are interested in independence and in getting to "the bottom line," yet this need is inconsistent with students' desires for change or improvement.

Hence, the research investigated several ways to bridge the gap between students' urge for independence and the limitations of their capacity for self-control. One method of bridging this gap was investigated in this study. Students in tenth grade were given a complex exercise. They were given the freedom and responsibility to use this exercise to prepare for a test. The results demonstrated higher motivation of students as well as higher achievement. The benefit was especially pronounced for students who were normally very quiet and reserved.

Teaching to the NCTM Standards

The NCTM Learning Principle states that a

> major goal of school mathematics programs is to create autonomous learners, and learning with understanding supports this goal. Students learn more and learn better when they can take control of their learning by defining their goals and monitoring their progress. When challenged with appropriately chosen tasks, students become confident in their ability to tackle difficult problems, eager to figure things out on their own, flexible in exploring mathematical ideas and trying alternative solution paths, and willing to persevere.[13]

Providing complex problems that afford students an opportunity to approach problem solving freely, unencumbered with any preconditions, makes them stakeholders in their learning and advances the goal of developing autonomous learners. In the classroom application below, students were given many choices in the design of their instruction and exercises. The research suggests that this is beneficial to student learning and furthers the goal of creating autonomous learners.

Classroom Applications

One method of giving students freedom with exercises is as follows. Plan a period of complex exercises where students have to apply varying knowledge as well as skills. Assignments in plane and solid geometry or systems of equations are very suitable for such complex exercises because they present a variety of possible ways of solving a problem. In this complex exercise students have a variety of requirements; they receive instruction and get help from the teacher. The basic idea is to distribute lessons and priorities of the exercise.

The lessons (time line):

Figure 3.6

1.	2. 3. 4. 5.	6.	7. 8. 9.	10.
LP	Exercises	Interim Balance	Individual	Final Test
ST	Basic Groups	ST	Consolidation	

LP = Lesson for Planning
ST = Short Test

In the lesson for planning (LP), the goal of the exercise is explained and students are allowed to work according to their own standard. In the tenth lesson (testing), every student selects four problems from a series of ten problems of different degrees of difficulty. Each of the students demonstrates in this way his or her willingness to make an effort and her or his learning progress. During the first exercise period (Lessons 2 to 5) most students are allowed to compose their own exercise program. At the end of a lesson, students have to make a note about problems and experiences they had.

The basic group was made up of students who were weak. They worked under direction of the teacher. In the sixth lesson, the teacher and students struck an interim balance (exchange of problems and experiences). Based on the results of the short test, the teacher's observation, and students' self-assessment, the second exercise period is used

1. to strengthen basic mathematical knowledge for students with problems
2. for individual consolidation by means of series of problems with increasing degrees of difficulty
3. for students who have finished their assignments to help others

The tenth lesson becomes a sort of "moment of truth."

Precautions and Possible Pitfalls

If you plan such a unit, try to draw up a three-phase period as in the example above: exercise, balance, and consolidation. In the study cited, other numbers of phases did not provide the success that three phases did and were not as manageable.

Source

Jürgen Westphal. Unterricht mit Jugendlichen—Das Lernen älterer Schüler (Lessons with youth—How older students learn). *Pädagogik und Schulalltag*, 46 Jg., Heft 2 (1991).

STRATEGY 46: Emphasize higher-level thinking objectives in regular mathematics classes so that all students incorporate the features of enriched academic and honors classes.

What the Research Says

A study was conducted in sixteen high schools with 1,205 classes taught by 303 teachers in California and Michigan. Grade levels ranged from nine through twelve. Teachers were asked to specify their instructional goals for each of their classes. The results showed that enriched classes emphasize high-level thinking objectives such as understanding the logical structure of mathematics, understanding the nature of proof, knowing mathematical principles and algorithms, and thinking about what a problem means and ways it might be solved. In contrast, regular mathematics classes tended to emphasize lower-level thinking objectives such as memorizing facts, rules, and steps; performing computations with speed and accuracy; and developing an awareness of the importance of mathematics in everyday life.

Teaching to the NCTM Standards

The NCTM *Research Companion to Principles and Standards for School Mathematics* summarizes the nature of teaching mathematics that is currently prevalent in the United States: "The most common pattern of classroom practice was extensive teacher-directed explanation and questioning followed by student seatwork on paper and pencil assignments."[14] Higher-level comprehension objectives were not observed in most mathematics classrooms. This is supported by the "video study" by the Third International Mathematics and Science Study (TIMSS), in which the typical eighth-grade mathematics lesson was

> organized around two phases: an acquisition phase and application phase. In the acquisition phase, the teacher demonstrates or leads a discussion on how to solve a sample problem. . . . In the application phase, students practice using the procedure by solving problems similar to the sample problem.[15]

The TIMSS video study showed that "teachers gave little attention to helping students develop conceptual ideas."[16] Enrichment topics can provide the "connection" that many students need to make sense of a problem and to provide a better understanding of why and how to follow a given procedure. Limiting instruction to procedural techniques does not support the NCTM learning standards or foster mathematical thinking.

Classroom Applications

Enrichment is not reserved for the gifted! There are many topics that are "off the beaten path" and can be made to suit an average ability class, or even a low ability class. The trick is to have the sensitivity to make these content adjustments. For example, students in lower level classes who are still struggling with numerical facts, or arithmetic algorithms, or perhaps with some simple algebraic skills, could benefit greatly by having teachers digress to consider the nature of parity of numbers, or to consider some arithmetic shortcuts that have some nifty mathematical justifications. These digressions exhibit a modicum of higher- order thinking skills. Such activities serve to enrich the instructional program both in the thinking skills provoked as well as the topics considered. Such digressions or enrichment ideas can be found in many sources. Some of these are under the rubric of "recreational mathematics" that can be found in books designed for mathematics teachers. One such is Posamentier and colleagues' (2006) *Teaching Secondary School Mathematics.*

Precautions and Possible Pitfalls

Perhaps the most important precaution when enriching a mathematics class (besides selecting appropriate material) is to be sure not to underestimate your students! Also, don't overemphasize the importance of the enrichment activity or it might render the regular instructional program less interesting. This would defeat the purpose of this enrichment activity: to motivate students toward further study in mathematics.

Sources

Posamentier, A. S., Smith, B. S., & Stepelman, J. (2006). *Teaching secondary school mathematics: Techniques and enrichment units* (7th ed.). Upper Saddle River, NJ: Prentice Hall.

Raudenbush, S., Rowan, B., & Cheong, Y. F. (1993). Higher-order instructional goals in secondary schools: Class, teacher, and school influences. *American Educational Research Journal, 30*(3), 523–553.

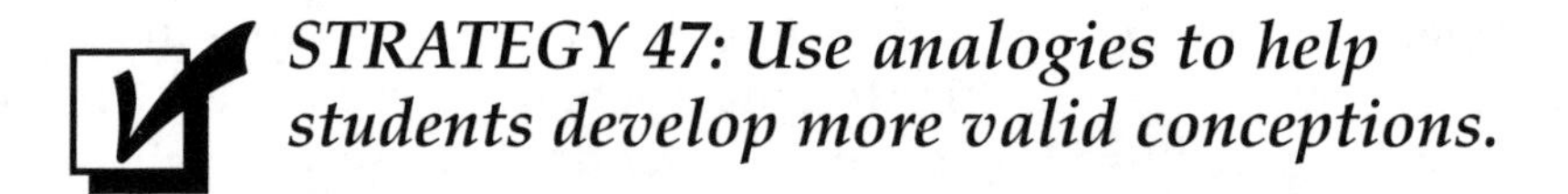

STRATEGY 47: Use analogies to help students develop more valid conceptions.

What the Research Says

Teachers' explanations of ideas by giving students analogies helps students construct more accurate conceptions of complex ideas. A study of 106 middle school students' (ages twelve–fourteen) ideas about electricity included studying students' conceptions of electrical circuits before their teacher covered the topic. Researchers examined students' use (and lack of use) of these preconceptions to explain how a light bulb and a battery work. Researchers also looked at how these ideas changed after instruction. What they found was that these ideas changed by becoming more internally consistent, but they did not become more valid. When the teacher used an abstract "thought experiment" analogy with a train on a closed-loop track to explain ideas about electrical circuits in a discussion, the students developed more accurate conceptions. Students were also able to use the analogy appropriately and to describe the limitations of the analogy. The chart below shows the analogy between the train and the electrical circuit (Joshua & Dupin, 1987, p. 131). The teacher also drew a diagram showing the train analogy on the blackboard and showed the connections to how an electrical circuit works.

Figure 3.7

Train	*Electrical Circuit*
Cars	Electrons
Movement of cars	Movement of electrons
Rate at which cars pass a certain point along the track	Rate at which electrons pass a given point in the circuit (current intensity)
Mechanical friction (obstacle in the track)	Electrical resistance (atomic nuclei)
Men pushing train	Battery
Muscular fatigue of men	Wearing out of the battery
Vibrations of the cars, noise and heat produced by collisions with obstacle on the track	Heat in the wires and battery, heat and light in the bulb produced by the interactions of the electrons and atomic nuclei

Teaching to the NCTM Standards

The NCTM *Principles and Standards for School Mathematics* encourages students to "make connections among mathematical ideas and to understand how mathematical ideas interconnect and build on one another to produce a coherent whole."[17] The classroom teacher is advised to lead students toward these connections by using a variety of tools, including making analogies that clarify and simplify some complex mathematical ideas. In addition, in the NCTM Professional Standards for Teaching Mathematics, teachers are encouraged to use "metaphors, analogies, and stories"[18] to enhance discourse as well as to use "concrete materials used as models"[19] to aid in the delivery of instruction. As the standard suggests and the application below supports, the students' role in discourse is to "try to convince themselves and one another of the validity of particular representations, solutions, conjectures, and answers."[20]

Classroom Applications

1. Do a concrete, hands-on classroom activity that involves the concept you are trying to teach, and follow it with a discussion.
2. Through the discussion, identify students' pre-instructional ideas about the concept.
3. Make sure students are aware of the different interpretations of the same concepts.
4. Have students evaluate the competing interpretations.
5. Give students a task that enables them to test out and decide between these interpretations. This will also help them to develop new interpretations and/or to refine old interpretations, as needed.
6. Have students draw conclusions based on this task. Help them identify and try to reconcile major contradictions.
7. Give students an analogy for the concept. Have them discuss the analogy, including its relevance and limitations.

In mathematics, finding analogies to use for developing more accurate conceptions may be more difficult because the material is more abstract and skill oriented. However, one possible example of where a model may simulate an abstract concept is to discuss the concept of a function using the model of a gun shooting bullets at a target. The bullets comprise the "domain" and the target comprises the "range." The gun, and its aiming, is the function. Because a bullet can be used only once, we know that the elements in the domain can be used only once. The gun can hit the same point on the target more than once, however, so points in the range can be used more than once. Some points on the target may never be hit even though all the bullets are used. Thus, a function is a mapping of all points of the domain to points in the range. When all the points on the target are hit, then the function is an "onto" function; when each point in the range is used only once, then we have a "1–1" function. We have a "1–1 onto" function when both properties are attained. This is also called a "1–1 correspondence." Using the gun-shooting-the-target analogy makes the concepts of function, domain, and range easier for students to understand and remember.

Precautions and Possible Pitfalls

Students often have difficulty fitting new ideas into preexisting frameworks. Make sure the preexisting framework is very clear before using it as an analogy. Students must have a clear understanding of the framework to be used in the analogy, such as a gun shooting bullets at a target, in order for the analogy to be a useful tool for understanding mathematical concepts such as function, domain, and range.

Source

Joshua, S., & Dupin, J. J. (1987). Taking into account student conceptions in instructional strategy: An example in physics. *Cognition and Instruction, 4*(2), 117–135.

4

Assessing Student Progress

STRATEGY 48: Feedback on practice is essential for improving student performance.

What the Research Says

Studies have shown that improved student performance is tied to the amount of feedback given to students. Students need to receive specific feedback on the results of their practice in order for learning to be effective. Practice with specific feedback results in more successful and more efficient learning.

Teaching to the NCTM Standards

The NCTM Assessment Standards for School Mathematics state that "continuous assessment of students' work not only facilitates their learning of mathematics but also enhances their confidence in what they understand and communicate."[1] The application below, which focuses on peer review or teacher review of students' homework, provides students with feedback on their work so "they can reflect on their progress, understand what they know and can do, be confident in their learning, and ascertain what they have to learn."[2] Eventually, we want students to "evaluate, reflect on, and improve their own work—that is, to become independent learners."[3]

Classroom Applications

In a mathematics instruction program, there are many opportunities for practice of skills presented. By pairing students and having them read each other's work, or by having students compare their work to model solutions, a form of feedback can be obtained regularly without a great expense of time. Teachers might also systematically review a small and different sampling of student papers each day, and from this small number of collected papers provide some meaningful feedback to the students. For example, supposing that a classroom is situated in rows of students. The teacher may "at random" call for the papers for everyone sitting in the first seat of each row, or from the students sitting on the diagonal, or from everyone in the third row. If a teacher wants to check on a particular student's paper a second day, as there may be some serious questions about the student's work, then the teacher can ask for the student's paper by including him or her in the second day's set of collected papers. This can be done by calling on the set that also describes that student's seat. For example, call on the third row one day and then, since the target student is sitting in the last seat of the row, call for the papers from all students sitting in the last seat of a row. This would "inadvertently" include the target student a second time.

Because it is unreasonable to expect the teacher to do a thorough reading of everyone's paper every day, there are alternative ways to give students feedback on their homework. One could search for parent volunteers and/or retired teachers who might like to take on some part-time work in reading and reacting (in writing) to student work. One might also try to engage some older and more advanced students to undertake a similar activity, using a "cross-age" tutoring approach. This would also serve the advanced students well as they can benefit by reviewing previously learned material from a more advanced standpoint. By doing this, not only are the target students being helped but the older student is deepening his or her knowledge of mathematics.

Precautions and Possible Pitfalls

Teachers often do not have sufficient resources to provide individual feedback to each student. When having students give each other feedback, teachers should be aware that the feedback from students will be of a different nature and certainly not a replacement for that provided by the teacher. Student feedback must be monitored to avoid perpetuating flawed ideas or misconceptions. The same holds true for teachers' aides, parent volunteers, or retirees assisting in the classroom.

Source

Benjamin, L. T., & Lowman, K. D. (Eds.). (1981). *Activities handbook for the teaching of psychology.* Washington, DC: American Psychological Association.

STRATEGY 49: Promptly give students information or feedback about their performance.

What the Research Says

Students need to know what they know and what they can do well. They also need to know what they do not know and what they cannot do well. Students often cannot make these evaluations on their own, so they need this kind of information from their teacher. Information about their knowledge and performance, which is known as feedback, can help students focus their learning efforts and activities. This helps students learn. Feedback is more meaningful and more useful when delivered in a timely fashion.

Teaching to the NCTM Standards

The NCTM Assessment Standards ask of teachers: "How does the assessment provide opportunities for students to evaluate, reflect on, and improve their own work—that is, to become independent learners?"[4] Providing informative feedback to students in a timely fashion is a major factor in having students become independent learners, capable of recognizing content areas that need strengthening. The classroom application below suggests that teachers should provide feedback that focuses on the depth of understanding and not simply check for right or wrong answers.

Classroom Applications

Homework in the mathematics classroom usually consists of problems with a very definite answer. It is very tempting for teachers to spot-check homework by inspecting to see if the right answers are offered without looking at the method to reach the answer. Whenever possible, teachers should thoroughly examine students' homework answers and methods; they should give students information about the quality of their performance.

Where a teacher's class is too large to do a thorough check of the homework, the teacher can select different subgroups from within the class daily, picking their homework from the collected class set. Teachers should carefully inspect the work, focusing more on the method than on merely seeing if the right answer had been obtained, and should provide timely, specific feedback.

Precautions and Possible Pitfalls

The teacher may either randomly select subgroups or select them by design. In any case, this selection should not be predictable by the students. Otherwise, those who anticipate homework inspection will do a better job and those who don't, won't. If feedback is not provided in a timely fashion, it will be of limited usefulness to students.

Source

Chickering, A., & Gamson, Z. (1987). Seven principles for good practice in undergraduate education [Special insert]. *The Wingspread Journal, 9*(2).

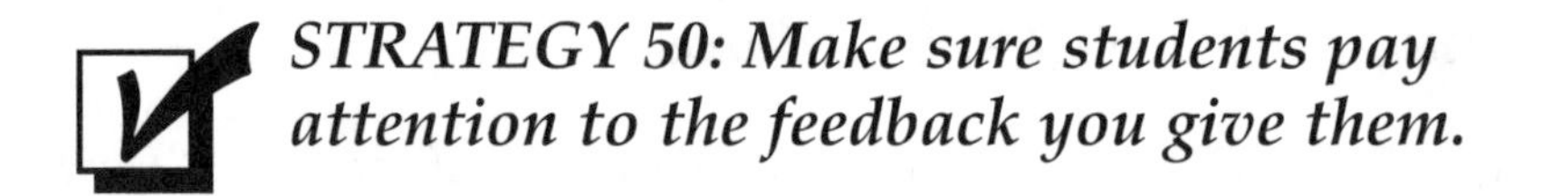

STRATEGY 50: Make sure students pay attention to the feedback you give them.

What the Research Says

Paying attention to feedback on items that were incorrect is related to achievement. There are two factors that affect whether students pay attention to feedback. One is whether students perceive that they can understand the teacher's feedback. The other is whether students focus on the negative feelings that arise from making mistakes. In a study of thirty-eight high school students in two classes, researchers observed how students processed feedback during computer programming lessons while the teacher discussed the results of a recent test. Observations were categorized into ten "on-task behaviors" (e.g., looking at the teacher or writing on the test) and nine "off-task behaviors" (e.g., looking out the window or writing on irrelevant material).

Thirteen low- and high-achieving students were randomly selected for interviews in order to get more detailed information on how they processed feedback. One distinct pattern that frequently emerged was students' judgment that they could not understand the teacher's feedback. When students do understand the feedback, they listen to what the teacher is saying and try to figure out what they did wrong. When they do not understand the teacher's feedback, they tune out. The other pattern that emerged, but was less common, was getting upset about making errors. When this occurred, instead of focusing on the problem, students tended to focus on their negative feelings.

Teaching to the NCTM Standards

The NCTM Professional Standards for Teaching Mathematics specifically define appropriate roles for students and teachers in mathematical discourse. The teacher must "decide when to provide information, when to clarify an issue, when to model, when to lead, and when to let a student struggle with a difficulty."[5] One of the student's roles in the discourse is "to listen to, and question the teacher."[6] Teacher feedback should be provided in a manner that is supportive and encourages the student to take action to apply the intended feedback toward improvement. Feedback that is intended solely to inform students that they don't understand a concept is not useful. However, feedback that provides a second opportunity for assessment (formal or informal) sends a strong message to students that the material is valued and that student mastery of it remains important. The application below highlights this principle as the student is given an opportunity to display comprehension through successful solution of a similar problem. Feedback and this second opportunity for assessment hold students accountable for the material covered in class in spite of any difficulties they may have encountered on prior assessments.

Classroom Applications

Perhaps the best way to ensure that a teacher's feedback is heeded is to have the students write about their error: what it was, why they made it, and how they would now solve the problem involved in the situation addressed by the teacher. For example, when a student could not do a proof of a particular geometric theorem, the teacher should expect the student to write about the correction and then demonstrate with another theorem that this problematical situation is now resolved. A student who just got some useful feedback from the teacher may also be asked to do a future and similar problem on the chalkboard and discuss it with the rest of the class. This will ensure that the student understands what the teacher has told him or her and does not gloss over the response just to get the teacher off his or her back. Another strategy is for teachers to have students keep a journal of their errors and regularly make journal entries when getting corrected papers returned or when reviewing work orally in class.

Precautions and Possible Pitfalls

Teachers should be aware of the fact that some of their comments, whether given individually or to the class, may be ignored or simply forgotten. Simple awareness of the importance of the students' retaining teacher feedback is already one big step in making this aspect of the instructional program effective. Journal entries and/or written error analyses can become tedious and should take on various forms. For example, the student might see this additional written assignment as a form of punishment. If this is sensed by the teacher, there should be an alternative way of reaching the same objective. In this case, the teacher may have the student who got the teacher feedback explain the problem and the teacher resolution to a classmate.

Source

Gagne, E. D., Crutcher, R. J., Anzelc, J., Geisman, C., Hoffman, V., Schutz, P., & Lizcano, L. (1987). The role of student processing of feedback in classroom achievement. *Cognition and Instruction, 4*(3), 167–186.

STRATEGY 51: Systematically incorporate review into your instructional plans, especially before beginning a new topic.

What the Research Says

Research on the types and timing of review in mathematics teaching shows that daily review of homework assignments is not enough. Review should be systematically integrated into lesson plans, especially before beginning a new topic. Such reviews help the teacher to plan so that students have the prerequisite knowledge and skills needed to successfully learn new material. Studies have shown that in addition to helping teachers plan instruction, review helps students (1) consolidate what they have learned, (2) summarize the main ideas, (3) develop generalizations, (4) develop a more comprehensive view of the topics, (5) get a "big picture" of how ideas fit together, and (6) feel confident that they are ready to move on to a new topic. Research on the timing of review suggests it is more effective when it is interspersed throughout the curriculum instead of being concentrated at one period of time. Research has been conducted on a range of review techniques. Studies show that student-generated outlines force students to organize ideas and structure the relationships between them. Such outlines have been found to enhance the recall of mathematical ideas. Review questions have been found to aid memory by increasing understanding. Questions can be word- or calculation-based. Research indicates that word-type questions require students to comprehend concepts and rules well enough to apply them to new situations. In contrast, calculation-type questions generally require understanding of only a small range of concepts and rules, and often involve only rote learning.

Teaching to the NCTM Standards

The NCTM Learning Principle states, "Students must learn mathematics with understanding, actively building new knowledge from experience and prior knowledge."[7] Prior knowledge may have been learned years earlier, so it is important for teachers to carefully create a skills inventory of material that students will need to know and to present review material before getting into the central themes of the lesson. This will allow for a lesson that should flow smoothly, with students focusing their attention on the new material and not being distracted by their inability to recall previously learned materials. In the NCTM *Handbook of Research on Mathematics Teaching and Learning,* it is stated that, "A mathematical idea or procedure or fact is understood if it is part of an internal network. The idea is that understanding in mathematics is making connections between ideas, facts, or procedures."[8]

Classroom Applications

Before beginning a new topic, teachers should review to identify which prerequisite knowledge and skills students have acquired, which should be taught again for reinforcement, and which are not yet known. When conducting a review, the teacher should include a broad range of content, from simple skills and concepts to the most difficult skills and concepts. There are several different types of review used in mathematics, including outlines, questions, homework, and tests.

Simply asking students if they remember a certain topic is insufficient for determining their readiness for a new topic. For example, when embarking on a fairly algebraically "heavy" topic in the midst of the geometry course, it would be wise to select the main skill that will be required of the students in that unit and give them a short informal quiz on the topic. If, for example, the Pythagorean Theorem is to be studied, then some review of topics such as radicals would be appropriate.

You might also give students a series of questions that should be done at home. Those would be designed to require simple ideas from the past. Students will then have a "pressure-free" opportunity to exhibit their knowledge after conducting a private review.

Precautions and Possible Pitfalls

When embarking on a review of things previously taught, avoid being dissuaded by students who are quick to tell the teacher, "We already had this stuff." But did they learn this material? This is what must be ascertained by the teacher. One effective way of doing this is by having students demonstrate and explain to the teacher what they know about the material to be reviewed rather than boring the students by repeating what they have been shown or told before.

Source

Suydam, M. N. (1984). *The role of review in mathematics instruction.* ERIC/SMEAC Mathematics Digest No. 2. ERIC Clearinghouse for Science, Mathematics, and Environmental Education, Columbus, OH. (ERIC Document Reproduction Service No. ED260891)

STRATEGY 52: Provide all students, especially students lacking confidence, with "formative assessments" to allow them additional opportunities to succeed in mathematics.

What the Research Says

Extensive research has shown that "formative assessment" has had significant impact on students' success in mathematics and science. In the King's-Medway-Oxfordshire Formative Assessment Project (KMOFAP) and a parallel project at Stanford University, nineteen teachers used formative assessments that provided enhanced feedback through their communications with students about homework, classwork, and tests. Those classes where students received feedback that was based on student learning and not on student accountability showed an average 0.3 standard deviation increase on standardized tests. These improvements, replicated throughout an entire school, would raise it from the lower quartile to well above average.

Teaching to the NCTM Standards

The NCTM Teaching Principle states that, "Effective mathematics teaching requires understanding what students know and need to learn and then challenging and supporting them to learn it well."[9] Students who have previously struggled in mathematics tend to lack confidence in their ability to succeed. Providing feedback that is useful to students is an important step in allowing them to identify their errors and make appropriate corrections. Affording students the opportunity to show the teacher that they have fixed an incorrect answer gives them an emotional boost and is educationally supportive. Thus, formative assessment that encourages additional effort and offers hints for successful problem solving builds confidence, increases students' self-esteem, and provides valuable information to both the students and the teacher about student progress.

Classroom Applications

Teachers can teach students how to convert failures into future successes by providing meaningful feedback and an opportunity for students to correct their mistakes. This "formative feedback" encourages and directs students to find the correct answer. Have students systematically analyze their mistakes to identify why they obtained the wrong answer. Have them highlight strategies to prevent similar errors in the future. Students should be able to explain precisely what they did wrong and how they can avoid similar mistakes in the future.

Teachers should ask:

1. What did you learn that you need to remember for future success?
2. What was the most difficult aspect of solving this problem?
3. How did we learn to overcome the difficult parts of this problem?
4. Where have we seen this type of problem before, and how did we deal with it?

Precautions and Pitfalls

While formative assessments provide an opportunity and roadmap for students to correct their errors, student accountability should not be diminished. Teachers must be certain to maintain an atmosphere where getting it right the first time is greatly preferred. All students should be expected to arrive at a correct answer, given additional hints and guidance.

Source

Black, P., Harrison, C., Lee, C., Marshall, B., & Wiliam, D. (2003). *Assessment for learning: Putting it into practice.* Buckingham, UK: Open University Press.

STRATEGY 53: Find out why students rate a mathematical task as difficult so you can increase the difficulty of exercises and tests more effectively.

What the Research Says

Systematically increasing the difficulty of exercises and tests helps reinforce students' motivation to learn and also produces better test results. However, research shows that teachers and researchers perceive difficulty very differently from students. Consequently, it is important for teachers to find out how their students feel about the difficulty of mathematical tasks so they can construct tasks with the appropriate increase in difficulty. A group of forty ninth-grade students (two classes) took a test consisting of four sets of eight arithmetical tasks of increasing difficulties. The test lasted thirty-five minutes. Every set formed one page. Three grades of difficulty were used: easy (*E*), medium (*M*), and difficult (*D*). Students were divided into Groups A and B. Their tasks were the same; however, the increasing difficulty of the tasks changed. Group A had to solve the three main sets in the sequence *EMD,* while Group B had to solve the three main sets in the sequence *MED.* At the end of every page, students estimated their success ("How many of the eight tasks do you think you solved correctly?"). In addition, students used a seven-stage scale to rate the difficulty of the set as well as the fun they had on each page and the success they expected for the next page. The results showed that Group A (*EMD*) had significantly more correct solutions than Group B (*MED*). Students in Group B rated the level of difficulty higher than students in Group A. Whereas teachers and researchers described tasks as medium difficulty when 50% are solved, students rated tasks as medium difficulty when they were 82% solved and rated them as difficult when 71% were solved! Students who thought they had correctly solved all eight tasks in a set rated most of the tasks as easy.

Teaching to the NCTM Standards

The NCTM Learning Principle states that, "Students must learn mathematics with understanding, actively building new knowledge from experience and prior knowledge."[10]

To properly assess what degree of learning has taken place, students must demonstrate that they have a firm grasp of procedural and conceptual knowledge of a topic. If the Learning Principle calls for the building of knowledge from experience and prior knowledge, the assessment must be constructed so that there is a "ramping" up of questions from easy to medium to hard. An examination that is not crafted in such a manner provides a psychological challenge that may interfere with the main goal of assessing a student's mathematical proficiency. Teachers can use a variety of methods to obtain this information; however, a careful analysis of student performance on homework questions can provide very useful information about the degree of difficulty of items. This information, augmented by student questionnaires and interviews, will give the teacher valuable insight into levels of difficulty for assessment purposes.

Classroom Applications

If teachers choose to increase the degree of difficulty, they need information about students' feelings and perceptions. To get this information, teachers can create a test and link it to questions that have students rate the difficulty of each item on a seven-stage scale such as the one in Figure 4.1.

Figure 4.1

1	2	3	4	5	6	7
Very easy	Pretty easy	Easy	Medium difficulty	Difficult	Pretty difficult	Very difficult

Similarly, teachers can investigate students' enjoyment and their expectations. Sometimes it is wise not to make assumptions about how students feel or to trust one's own theories or sixth sense about students' reactions. Instead, teachers should ask the students themselves. One way to do this is through interviewing students. Another way is to administer a questionnaire and obtain statistical data on students' feelings about tasks, which provides objective information that can help the teacher design tasks consistent with students' abilities as well as their motivation. Students might also appreciate having their feelings about academic tasks examined.

Precautions and Possible Pitfalls

Do not ask students about their feelings about task difficulty and enjoyment too often. Otherwise, you are likely to get a distorted impression of the students' estimation of an assignment's difficulty, and students may deceive you in order to make exercises and/or tests easier!

Source

Marion Kloep & Franz Weimann. Aufgabenschwierigkeit und Mathematikleistungen bei Realschülern. Zum Problem der mittleren Aufgabenschwierigkeit (Student's task difficulty and achievement in mathematics. On problems of "medium task difficulty"). *Psychologie in Erziehung und Unterricht,* 29 Jg., (1982) S. 76–80.

STRATEGY 54: Increase your understanding of factors that affect students' attitudes before and after testing. You may be surprised!

What the Research Says

Students fear tests less than usually is assumed. The widespread negative attitude of many students before a test is mainly due to their competence in the content they are being tested on rather than the fact that they are being tested. Research has shown that students' attitudes did not deteriorate, as was predicted, when they took a test. A study was conducted to investigate specific emotional states of students in different school situations. Four attitudinal dimensions were examined: negative attitude (depression, anxiety, boredom, tiredness, and nervousness); positive attitude (good mood, activity, carelessness, and relaxation); eagerness to learn (concentration, sympathy); and aggressive excitement (aggressiveness, annoyance). Participants were 126 students; seventy-two were boys and fifty-four were girls. The average age was eleven years and six months. Students judged fifty-eight items using a six-point scale. The items asked about their emotions before and after a test and about lessons with and without testing. The researchers hypothesized that (1) all four dimensions would show differences before and after a test, (2) students would show different attitudes with and without a test, (3) girls and boys would show different attitudes, and (4) students' attitudes would depend on their grades.

The results showed that (1) only two of the four dimensions showed differences before and after the test; they were positive attitude and eagerness to learn; (2) in general, both girls and boys had better attitudes after a lesson; (3) girls showed less eagerness to learn than boys; (4) taking the test did not produce the expected deterioration of students' emotional state; (5) girls' emotions did not differ before and after lessons or with and without tests; in both cases girls had more negative attitudes than boys; and (6) students who got good grades showed better attitudes before the tests than students who did not get good grades.

Teaching to the NCTM Standards

The NCTM Assessment Principle states that "Assessment should support the learning of important mathematics and furnish useful information to both teachers and students."[11] Experienced teachers will tell you that a good test is a learning tool. Students who are assessed through a thoughtfully designed examination are appropriately rewarded for their ability to understand, reason, and apply concepts. The aforementioned research states that the act of taking the test did not cause the students' emotional state to deteriorate; this supports the notion that students are eager to demonstrate their understanding and identify their weaknesses. Administering formative assessments allows teachers to provide useful information to students that can improve student attitudes and confidence, and that ultimately leads to better performance. Teachers are reminded that assessments must be designed in many different ways to allow for every type of learner to demonstrate mastery of the content. In addition, students should be given an opportunity to correct any mistakes to improve their confidence and attitude, and to inform the teacher that the content has been mastered and the student is prepared to move ahead.

Classroom Applications

Teachers tend to notice this difference in attitudes of girls and boys when tests are given, but they don't realize that girls in general have a worse attitude than boys. The research suggests that students' attitudes depend upon their capability more than their anxiety about a test. A student's negative attitude both before and after a test can be a signal that it is necessary to support test results with careful comments (see Strategy 56, about students' perception of marks). If you provide students with feedback that they can use to improve their performance in the area being tested, and they start getting better grades because they have learned from their mistakes, then you can expect students to show more positive attitudes, regardless of testing. If you notice that students who usually have positive attitudes suddenly have negative attitudes, this observation can be a sign that students are having difficulty learning the content you are teaching.

Precautions and Possible Pitfalls

Do not consider every significant change in a student's attitude as an expression of a learning problem. There are many personal factors that can cause changes in students' emotional states.

Source

G. Chemnitz. Emotionale Reaktion von Schülern während einer Schulstunde mit und ohne Klassenarbeit. (Emotional reactions of pupils during school lessons with and without tests). *Psychologie in Erziehung und Unterricht,* 26 Jg., (1979) S. 170–173.

STRATEGY 55: Be aware of students' different levels of text anxiety as it relates to different subject areas, and use a variety of techniques to help them overcome their test anxiety.

What the Research Says

Students have different degrees of test anxiety for different subject areas. One study compared 196 first-year college students' self-reports of test anxiety in mathematics, physical sciences, English, and social studies. Students were administered the Worry-Emotionality scale in which they rated their anxiety about tests in one of these four subjects. The directions asked them to imagine they were taking a test in mathematics, for example, and to rate their feelings on a five-point scale ranging from "I would not feel that way at all" (1) to "I would feel that way very strongly" (5). Questions included, "I would feel my heart beating fast" and "I would feel that I should have studied more for that test." In rank order from most test anxiety to least test anxiety, the subjects were: physical sciences, mathematics, English, and social studies. For elementary and secondary school students, test anxiety is often developed from a combination of factors. These factors include parents' early reactions to their children's poor test performance, students' comparisons of their performance with other students as well as their own prior test performance, and increasingly strict evaluation practices as students progress through school. For low-achieving students, failure experiences tend to increase test anxiety. For high-achieving students, unrealistically high self, parental, and peer expectations tend to increase test anxiety. Some classroom practices affect test anxiety. Presenting material in an organized way and making sure it isn't too hard tends to improve the performance of test-anxious students.

One study (Bangert-Drowns, Kulik, Kulik, & Morgan, 1991) has shown that feedback is most effective when it is provided in a supportive manner, with an emphasis on guiding students to modifying their answer.

Teaching to the NCTM Standards

The NCTM Assessment Principle states that, "Assessment should support the learning of important mathematics and furnish useful information to both teachers and students."[12] Students who suffer from mathematical anxiety do not accurately demonstrate their progress on classroom assessments. The NCTM encourages "using multiple forms of evidence"[13] to assess student learning:

> For example, while multiple choice examinations can assess a broad range of content, open-ended assessments can probe more deeply into students' strategies and understanding. Long-term projects allow students to explore mathematical ideas in depth, and portfolios may help them to reflect on their best work.[14]

While the root cause of mathematical anxiety is traced to limited prior success, students who exhibit mathematical anxiety are less likely to continue to study mathematics and thus limit their career choices. Teachers can use test anxiety as a motivational tool to encourage students to study hard, be prepared, and ultimately be successful. Turning a student who suffers from mathematical anxiety into a confident, competent mathematician will provide a psychological boost that will benefit the student in the mathematics classroom and in life.

Classroom Applications

Test anxiety interferes with test performance. Students waste mental energy on anxiety that they could be using to answer the test questions. Make students understand that the test is one of many ways that they are being assessed. Explain to them that every homework assignment is a test. In fact, homework is an untimed, open-book test! Encourage them to participate in class as this is another way to verify their progress in learning mathematics. Once teachers have placed the classroom test in proper perspective, there are many strategies teachers can suggest and demonstrate to students to help them relax. First find out what strategies students already use. Share with them techniques you use to relax. Try to reduce the pressure students feel from being evaluated by tests. Using other assessment strategies, in addition to tests, can help reduce the pressure of being evaluated solely by tests and the corresponding test anxiety. Help students learn to differentiate between constructive and destructive instances of anxiety. In constructive or facilitative anxiety, students see tests as challenging experiences. In destructive or debilitative anxiety, students see tests as negative self-evaluation experiences. Teach students to become aware of and control their anxiety before, during, and after testing. For example, ask them, "What thoughts go through your mind before taking a test? What kinds of thoughts do you have while you are taking a test?" Help students improve their study strategies and test-taking skills. Demonstrate and encourage use of assorted relaxation techniques as described below.

Deep Breathing

With erect posture, breathe in deeply through the nose and hold your breath for a count of eight to ten. Then, slowly exhale through the mouth, counting to eight or ten. Repeat this procedure several times until relaxation occurs.

Muscle Relaxation

a. *Tension-Relaxation.* Tighten and then relax a muscle or set of muscles, like your shoulders, that normally store considerable tension. Hold the muscles in a tensed state for a few seconds and then let go. Repeat this sequence with the same muscles a few times and then move on to other muscles.

b. *Self-Hypnosis.* Sit straight in a chair with arms and legs uncrossed, feet flat on the floor, and palms on top of your thighs. Progressively relax your body, from toes to head, systematically focusing on one part at a time. Concentrate on tuning in to your bodily sensations, allowing your muscles to relax, becoming more aware of what it feels like when your muscles are relaxed. Talk to yourself (aloud or silently), telling yourself to loosen up and lessen the tightness. When the body is relaxed, it is more receptive to positive self-talk. Build up your self-confidence at this point. For example, "I know I can do well on this test!"

Creative Visualization

Guide students in engaging in success imagery weeks, days, and minutes before testing.

a. *Olympic Success.* Tell them to try what the Olympic athletes do to develop confidence in their performance. Picture yourself in a tense situation, like taking a test, and visualize yourself looking over the test, seeing the questions, and feeling secure about the answers. Imagine yourself answering the questions without too much difficulty. Complete the picture by imagining yourself turning in the paper and leaving the room assured that you did your best.

b. *Relaxing Place.* Where do your students feel most at peace? One spot could be at the ocean. Have students identify a place and use all of their senses to imagine themselves there and how they feel when they are there. Guide them in an activity such as: "Watch" the waves with their white caps rolling up the shoreline onto the beach. "Listen" to the waves and the seagulls. "Smell" the salty air, and "feel" your fingers and toes in the warm, soft, and grainy sand.

Precautions and Possible Pitfalls

Whatever a teacher decides to do in this regard must be done with a modicum of reserve, assessing the audience and their reaction to the above. Not all suggestions work with all students. Make sure to encourage students not to give up if the first relaxation technique doesn't work. Often these techniques need to be practiced to be successful, and students frequently must experiment with a variety of techniques to determine which ones work best for them. Teachers should ask students questions to see whether gender and/or cultural differences might affect the use of the above suggestions and to elicit ideas not previously considered by the teacher.

Sources

Bangert-Downs, R. L., Kulik, C. C., Kulik, J. A., & Morgan, M. (1991). The instructional effect of feedback in test-like events. *Review of Educational Research, 61,* 213-238.

Everson, H. T., Tobias, S., Hartman, H., & Gourgey, A. (1993). Test anxiety and the curriculum: The subject matters. *Anxiety, Stress and Coping, 6,* 1–8.

Hartman, H. (1997). *Human learning and instruction.* New York: City College of City University of New York.

Wigfield, A., & Eccles, J. S. (1989). Test anxiety in elementary and secondary school students. *Educational Psychologist,* 24(2), 159.

STRATEGY 56: Do not assume that students accept responsibility for or agree with their bad grades on tests.

What the Research Says

Teachers should be aware that students often just partially accept a teacher's marks. Many students overestimate their achievements and expect better grades than they get. When giving reasons for their good and bad grades on reports and tests, students explained their good grades in terms of the efforts they made. In contrast, when explaining why they got bad grades, students attributed their bad grades on tests to "tough luck," and they explained their bad grades on reports to their "lack of efforts." A study was conducted with 146 students ages fifteen to eighteen. Students were asked to evaluate the responsibility for their grades in terms of four sources:

a. students themselves
b. teacher
c. other students
d. the situation

Students also rated their satisfaction with their grades and their perception of the fairness of the grade on a four-point scale. Satisfaction was measured by having them respond to the statement, "Because of my achievements I deserved a better mark." Fairness was measured by having them respond to "In comparison to the other students, my mark is not fair." The results showed that students' expectations for their grades were consistent with the teacher's 66% of the time, while 23% of the time they expected a better grade and 11% of the time they expected a worse grade. When students received grades that were lower than they expected, they were dissatisfied with their grades and judged them to be unfair. Finally, the results showed that students usually ascribed responsibility for their grades to someone other than themselves.

Teaching to the NCTM Standards

The NCTM Assessment Standards state that

> assessment contributes significantly to all students' learning. Because students learn mathematics while being assessed, assessments are learning opportunities as well as opportunities for students to demonstrate what they know and can do. Moreover, assessments, including those external to the classroom, guide subsequent instruction and thus they can further enhance students' learning. Students can also themselves use assessments to become independent learners. They can do so by using assessments as indicators of the mathematics important for them to learn. It is through our assessment that we communicate most clearly to students which activities and learning outcomes we value.[15]

Because assessments serve such an important function in the teaching and learning of mathematics, it is important that the feedback on the assessment contain specific comments that students can use to improve their understanding of mathematical concepts. It is also advisable for teachers to give students second opportunities to display comprehension of items whose mastery was not evidenced by the primary assessment. If you value the content, it is important to send a message to students that you want them to master it, even on a second assessment. The NCTM assessment standard asks: "How does the assessment provide opportunities for students to evaluate, reflect on, and improve their own work—that is, to become independent learners."[16]

Classroom Applications

Just as it is important for a teacher to praise or make specific comments about a student's good grades, the teacher should make specific comments about the reason for a student's poor grades. Otherwise the student is likely to externalize responsibility for his or her poor grades, explaining them in terms of too little practice, too difficult tests, not enough time, and so forth. The teacher should give detailed feedback regarding a student's weaknesses, yet in a constructive manner. Mathematics teachers sometimes tend to forget to do this. They expect knowledge that an answer is wrong or right to be sufficient. Far from it! Students who get poor grades attempt to justify their bad marks with external circumstances in order to preserve their self-respect.

Precautions and Possible Pitfalls

Don't embarrass or humiliate your students! To help preserve their self-respect it is recommended that you give students feedback on their poor grades in private—not in front of the whole class. Written notes or personal comments after the lesson are much more helpful. Recognize that students who have repeated learning weaknesses are likely to expect better marks than they get. Consequently, such a student's reaction is usually one of disappointment and the student is likely to be sensitive about it. In such situations many students tend to feign indifference or amusement about bad marks. Do not be discouraged by this or take it personally. Continue to provide feedback constructively about the reasons for bad grades.

Source

Henning Allmer. Selbstverantwortlichkeit und Schülerzufriedenheit nach erwarteter und unerwarteter Leistungsbewertung (Personal responsibility and pupil satisfaction after expected and unexpected performance evaluation). *Psychologie in Erziehung und Unterricht,* 29 Jg., (1982) S. 321–327.

STRATEGY 57: If students do not follow your instructions and/or if their achievements do not fulfill your expectations, the cause may not be students' incompetence. It could be a result of your self-overestimation.

What the Research Says

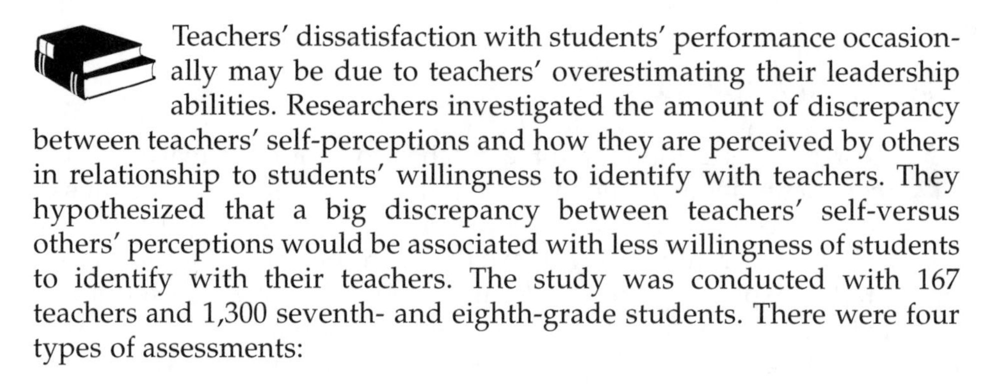

Teachers' dissatisfaction with students' performance occasionally may be due to teachers' overestimating their leadership abilities. Researchers investigated the amount of discrepancy between teachers' self-perceptions and how they are perceived by others in relationship to students' willingness to identify with teachers. They hypothesized that a big discrepancy between teachers' self-versus others' perceptions would be associated with less willingness of students to identify with their teachers. The study was conducted with 167 teachers and 1,300 seventh- and eighth-grade students. There were four types of assessments:

1. Heads of the schools where the teachers used to work evaluated teachers' leadership by scaling sixty prescribed items. This was the "outside assessment," independent of students' feelings about the teachers.
2. Students completed a questionnaire asking for their impression of a teacher's leadership.
3. Students completed a questionnaire asking for their willingness to identify with their teachers in two different ways: on a social-personal level and on a performance-oriented level.
4. Teachers completed a questionnaire asking for their self-assessment of their leadership abilities.

The results showed the following:

- Students have shown a strong tendency to identify with the teachers who underestimate their capacity to make high demands.
- Teacher's self-overestimation (i.e., self-assessment vs. outside assessment) makes students less willing to identify with teachers on the performance-oriented level and leads to personal rejection of the teacher.
- Student willingness to identify with teachers on the performance-oriented level is facilitated by teacher characteristics of patience, humor, and emotional safety.

- Student willingness to identify with teachers on the social-personal level is facilitated by teacher friendliness.
- Student willingness to identify with teachers on both levels increases when there is a high level of agreement between teacher's self-assessment and outside assessment or when the teacher's self-assessment is lower than the outside assessment.

Teaching to the NCTM Standards

The NCTM Professional Standards focus on key elements of the learning environment, particularly the role of the teacher as a classroom leader:

> What teachers convey about the value and sense of students' ideas affects students' mathematical dispositions in the classroom. Students are more likely to take risks in proposing their conjectures, strategies, and solutions in an environment in which the teacher respects students' ideas, whether conventional or nonstandard, whether valid or invalid. Teachers convey this kind of respect by probing students' thinking, by showing interest in understanding students' approaches and ideas, and by refraining from ridiculing students.[17]

The application suggests that teachers should focus on their behavior and not only focus on student shortcomings. Teaching is a communication art. Student failure to achieve levels of expectation should cause an introspective review of one's teaching practices.

Classroom Applications

Check the development of the above-mentioned characteristics (patience, humor, emotional safety, and friendliness) in your personality. Ask friends. Inspect your capacity to make high demands. Be introspective. If you feel that you lower your demands to match students' willingness to complete assignments, try changing your demands for a few weeks and look at the results. Consider the following possibilities for challenging students more:

- Make homework assignments more challenging
- Shorten the time allotted for written exercises
- Make students use correct grammar and vocabulary in their oral and/or written answers
- Examine students' notebooks

Consider having students work on an independent project resulting in a paper or report on a topic that is not part of the curriculum.

Precautions and Possible Pitfalls

Do *not* start with changes in grading students' work. They would see it as unjust. Students are very sensitive about being treated unfairly. Consequently, a sudden change in your grading practices would backfire, making them more unwilling to identify with you on a social-personal level and on a performance-oriented level. Gradually, in a step-by-step fashion, give them the new orientation to your demands on both levels. After three or four weeks students are likely to accept the new standards.

Source

Wolfgang Kessel. Selbstbild-Fremdbild—Differenz der Lehrer und Identifikationsbereitschaft der Schüler. (Difference between teacher's self-assessment and outside-assessment and the coherence with the student's willingness for identification). *Pädagogische Forschung,* 22 Jg., Heft 1 (1981), S. 533–560.

5

Teaching Problem Solving

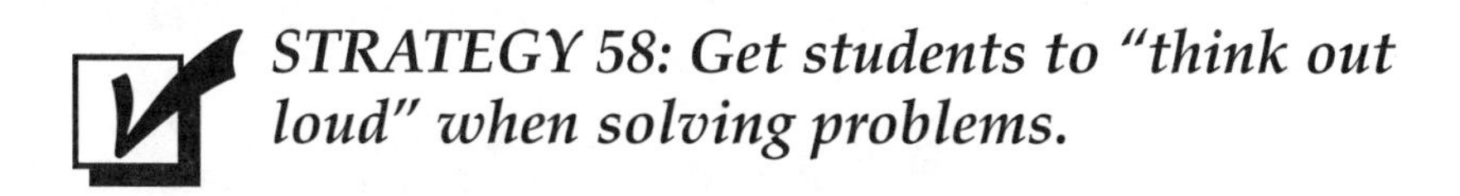

STRATEGY 58: Get students to "think out loud" when solving problems.

What the Research Says

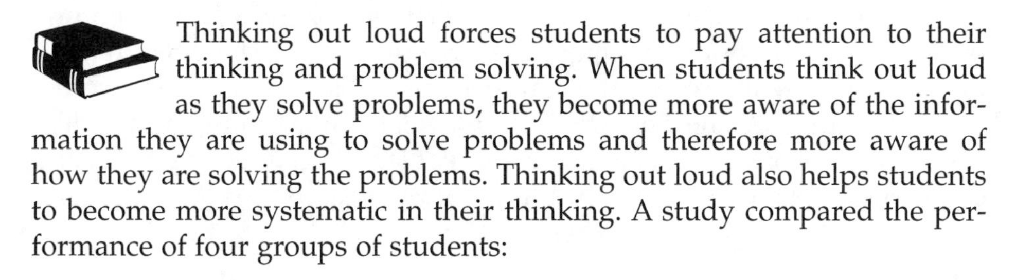

Thinking out loud forces students to pay attention to their thinking and problem solving. When students think out loud as they solve problems, they become more aware of the information they are using to solve problems and therefore more aware of how they are solving the problems. Thinking out loud also helps students to become more systematic in their thinking. A study compared the performance of four groups of students:

- Students who thought out loud with a specific mathematical problem-solving goal
- Students who thought out loud with a nonspecific mathematical problem-solving goal
- Students who thought silently with a specific goal
- Students who thought silently with a nonspecific goal

The think-out-loud groups were told, "Tell me, what equation are you going to look at or use, and why are you going to look at or use it?" The think-silently groups were told, "Think about what equation you are going to look at or use, and why you are going to look at or use it." All groups were given a series of numeral conversion problems to solve with bases of 3, 4, 5, and 10, using four equations. Two days later they were given a written test designed to measure what they had learned during the problem-solving session. The results showed that students who thought out loud solved problems more efficiently in terms of the number of unnecessary steps and the sequence of steps, but required more time to solve problems.

Teaching to the NCTM Standards

The NCTM Communication Standard calls for students to

- Organize and consolidate their mathematical thinking through communication;
- Communicate their mathematic thinking coherently and clearly to peers, teachers, and others;
- Use the language of mathematics to express mathematical ideas precisely.[1]

Asking students to think out loud encourages them to organize their thoughts and think logically. The application outlined below, having students work in pairs, gives each student the task of expressing, with mathematical clarity, the problem-solving strategies that they would employ in crafting a solution while the "listener" has the task of critiquing the approach and offering a more effective technique. This nonthreatening exchange will lead to the loftier goal of having students express their conjectures to the class and provide supporting evidence in a clear and concise manner. In this manner, student learning, understanding, and comprehension are increased through communication.

Classroom Applications

1. Have students work in pairs, with one thinking out loud while solving problems and the other listening analytically to how the problem is being solved. Rotate roles so all students serve as thinkers and listeners.
2. Think out loud for students as you solve problems to model for them how they should think about and solve problems. Purposely make mistakes so you can show students how to recover from errors.

NOTE: Let students know that smart people—not just crazy people—talk out loud to themselves!

Precautions and Possible Pitfalls

1. People usually think faster than they speak, so sometimes thoughts trip over speech.
2. If students don't have adequate knowledge about the problem, they may not be able to think out loud about how they would solve it.
3. Students might be shy about thinking out loud because of cultural backgrounds, speech impediments, or simply peer pressure. Be aware of this issue.

Sources

Whimbey, A., & Lochhead, J. (1982). *Problem solving and comprehension.* Philadelphia: Franklin Institute Press.

Zook, K. B., & DiVesta, F. J. (1989). Effects of overt, controlled verbalization and goal-specific search on acquisition of procedural knowledge in problem solving. *Journal of Educational Psychology, 81*(2), 220–225.

STRATEGY 59: Have students study written model solutions to problems while learning and practicing problem solving.

What the Research Says

Research was conducted in which one group of high school algebra students was taught to translate words into algebraic terms by studying written model solutions (worked-out examples) of how to make the translations and then given some practice problems. Students in this group studied a worksheet on writing equations that included three examples and three practice problems. After discussing the worksheet and sample problems, students completed three practice problems. After discussion of these three problems, students were given twenty-four additional practice problems, twelve of which were preceded by worked-out examples. Students in the other group did not study worked-out examples to problems but were given a greater number of practice problems to solve. Students in the written model solutions group were reminded to study the models and were given additional help using them if they needed it. Students in the practice-problems group were asked to think about the initial problems or a similar problem they had solved before. These students also received more direct guidance in understanding the problems and completing the algebraic terms from the teacher. After both groups went through these procedures, all students were given an eight-item test. After the test, all students were given an additional twenty problems for homework. Following the homework, all students were given a twelve-item test. No worked-out examples were given for either test.

The results showed that students in the written worked-out examples group made fewer errors on both tests; they completed their problems faster and with less help than students in the practice-only group.

Teaching to the NCTM Standards

The NCTM Professional Teaching Standards encourage the use of tools to enhance the understanding of mathematics. Included in this "toolbox" are concrete materials, written hypotheses, explanations, and arguments. Model problems that serve as a roadmap for the solution of similar problems should be part of the student's repertoire of strategies in problem solving. The application below describes how this practice can be applied to enhance comprehension and improve students' problem-solving skills.

Classroom Applications

Written model solutions to problems can be used as supports for homework, when doing individual instruction with remedial students, and in regular classroom instruction. Teachers should help students understand why and how a particular model fits a particular problem. Students ought to use worked-out examples with comprehension rather than employ rote memorization. Exercises that provide practice with algebraic skills lend themselves quite well to this scheme. Because the skills tend to be repetitive and similar, referring back to the model solution will be an important form of review and reinforcement of procedures. Although this sort of recollection is not considered real problem solving, it does provide some students with the necessary support to learn mathematical skills. It should be noted that real problem solving is also dependent to some extent on recalling previous problem-solving experiences. These sometimes manifest themselves as cleverly solved "model solutions." Thus, models, albeit at a higher level, similarly serve to provide a positive experience for mathematics students. There is a school of thought that believes that much of what one does in school mathematics is recalling prior examples of the same or similar situations.

Precautions and Possible Pitfalls

Using model solutions for a particular type of verbal problem can also be counterproductive in that it could set students up for solving any problem that looks like the model in the same way as the model. In real life, problems can subtly differ, and if a student is too dependent on seeing a model solution beforehand, it could diminish his or her creativeness in problem solving.

Source

Carroll, W. (1994). Using worked-out examples as an instructional support in the algebra classroom. *Journal of Educational Psychology, 86*(3), 360–367.

STRATEGY 60: Encourage students to make mental pictures while applying rules to solve problems.

What the Research Says

Creating and using mental images about rules can help students solve problems. A study was conducted with fifty-two ninth-grade students who were assigned to one of two groups. Both groups were given information on Boyle's Law, Charles' Law, and Gay-Lussacs law. One group of students was encouraged to create mental images of a typical gas as it responded to different amounts of pressure, temperature, or volume. This group was also instructed to draw an image of this in their notebooks. Students in the other group were not instructed to make mental pictures or draw images in their notebooks. These students were told to write the rule and repeat it out loud while learning. After three days of instruction on the gas laws, students were given an exam and completed a questionnaire designed to assess their use of imagery while they were taking the exam. The exam consisted of two parts. The first part was made up of six multiple-choice items designed to measure students' memory. The second part consisted of six essay questions in which students had to use the rules to solve problems. Students had to solve these problems correctly and give acceptable explanations for their answers.

The results showed that the imagery group did better than the nonimagery group in solving the problems, but the nonimagery group did better than the imagery group on the memory part of the exam. However, many other studies have demonstrated mental imagery can have a very beneficial impact on memory.

Teaching to the NCTM Standards

The NCTM Professional Teaching Standards of 1991 state, "The teacher of mathematics should engage in ongoing analysis of teaching and learning by examining effects of the tasks, discourse, and learning environment on students' mathematical knowledge, skills, and dispositions."[2] This suggests that teachers must be very sensitive to the various types of learning tasks in which their students are engaged. Because the use of imagery was profoundly effective for students applying rules to conceptual ideas, it should be employed as a teaching strategy in such cases. Similarly, repetitive activities effectively supported the type of learning that required memorization. While repetitive classroom activities that support memorization are typically discouraged, there are times when they are both necessary and advisable. Teachers should use multiple strategies to support the different learning tasks of their students.

The NCTM Problem Solving Standard calls for students to "apply and adapt a variety of appropriate strategies to solve problems."[3] Thus, procedural knowledge should be strengthened with conceptual understanding that is enhanced by visualization and imagery. The applications below are wonderful examples of employing simple visualization techniques to simplify complex problems. Encouraging students to use a variety of strategies to solve problems is consistent with the Standards and is a good practice in their daily lives.

Classroom Applications

The above research has clear implications for mathematics instruction. For learning algorithms, it makes sense to have students remember the rule being used. Perhaps later, a more in-depth study of the algorithm can be helpful. For the solution of problems that, by their very nature, do not call for a diagram, it is sometimes quite helpful to visualize the situation being described. Visualization can be in the form of sketches, diagrams, or mental pictures. A "mental picture" is a thought diagram. Consider the following problem:

> In a room of 45 children, 28 wear eyeglasses, 30 are wearing a white shirt, and 5 are not wearing a white shirt and do not wear eyeglasses. How many are wearing a white shirt and have eyeglasses on?

To do this problem most expeditiously, it would be wise to mentally picture the students and draw a Venn diagram of the situation.

Figure 5.1

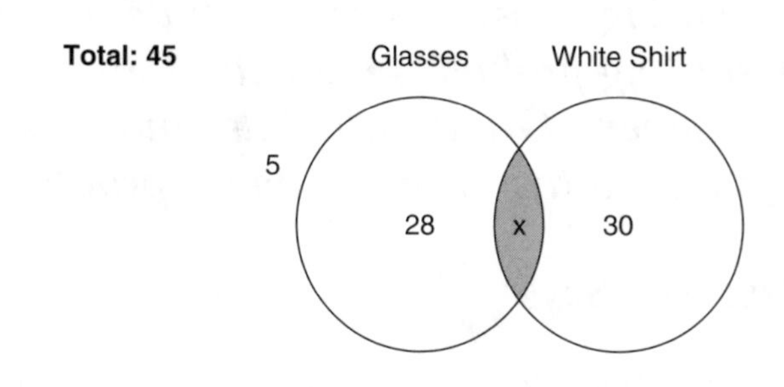

The intersection x can be found by the equation: $(28 - x) + (30 - x) + x + 5 = 45$, arrived at by adding the contents of each of the regions, which total 45. Therefore $x = 18$. Without drawing the diagram the problem would have been considerably more difficult. Nothing in the statement of the problem told the students to make the drawing. This is where the teacher's role is important. Teachers should directly instruct students in making drawings and/or mental pictures of problems they are solving. Students need to be encouraged and reminded to make and use mental images and diagrams.

In addition to drawing, students can form mental pictures of problems just by using their visualization skills. For example, when reading a problem like, "Alicia is taller than Eama, and Michael is taller than Alicia. Who is the tallest?" a student can make mental pictures of Alicia, Eama, and Michael to compare their relative heights.

Precautions and Possible Pitfalls

Drawing pictures to represent abstract concepts or topics is useful but is by no means the solution to all problems. Teachers should be careful to make this point clear. Yet, they should always encourage students to try to represent the problem situation graphically to gain more insight. Many students view mental images as "crutches" or forms of cheating and feel guilty about using them. Let students know it's smart to use such strategies.

Source

McIntosh, W. (1986). The effect of imagery generation on science rule learning. *Journal of Research in Science Teaching, 23*(1), 1–9.

STRATEGY 61: Provide hints or clues or ask leading questions when students need help solving problems instead of giving them the answers. Gradually phase out this support so as to foster independent problem solving.

What the Research Says

The instructional technique of scaffolding is based on a concept known as the "zone of proximal development." This zone is the distance between what students can do when solving problems independently and what they can do when solving problems with someone at a higher level. Studies have shown that students who cannot solve problems on their own can often solve them if they are given temporary supports or "scaffolds" by another person who is more competent in the particular area. This person can be a teacher, a parent, or another student. There are many different types of scaffolds that can be used, including hints, question prompts, partial solutions, and model solutions or worked-out examples. Working in groups to solve problems cooperatively also provides scaffolding opportunities. These forms of support act as bridges between what students can do on their own and what they can do with guidance. As students make progress, these forms of support are gradually withdrawn. The result is that, eventually, students can solve the problems on their own.

Teaching to the NCTM Standards

The NCTM Learning Principle states, "Students must learn mathematics with understanding, actively building new knowledge from experience and prior knowledge."[4] Effective teacher questioning helps students narrow the "zone of proximal development." Thus, teachers should lead students through difficult solutions through a series of well-constructed questions that emphasize the process over the procedure. Students should soon be able to be autonomous learners, capable of asking themselves the questions that will lead them toward crafting solutions of their own design.

Classroom Applications

The problems that arise when doing proofs in geometry often stem from the fact that students are more concerned with mimicking the teacher's model solution than with real comprehension. Through regular questioning, the teacher can systematically guide students through the proper "backwards method" for approaching a proof. These frequent and simple questions will eventually become part of the students' thinking process when doing proofs. Some sample questions (naturally, each teacher uses his or her own words) are: What are we trying to prove? What are some of the ways we can establish this idea? How can we reach this previous item? What are we given? Have we ever done a proof similar to this? And so on. Teachers should gradually reduce (and ultimately phase out) their questions as students begin to ask themselves the same sort of questions. To make sure this happens, follow these steps:

1. Model self-questioning when problem solving by thinking aloud for students as you solve different types of problems. Make sure they see how you tailor the questions to the specific problem.
2. Give students guided practice in self-questioning while solving different types of problems and give them feedback on their question formulation and use. Make sure they generate their own questions instead of simply copying yours.
3. Have students independently practice self-questioning while solving a variety of problems, but have them write down their questions at each stage of the problem-solving process so you can give them feedback.
4. Remind students to continue to ask themselves questions while problem solving on their own, even though they won't be turning in the questions for your feedback.

After these four stages, students will eventually ask themselves questions while solving problems automatically, without having to think about it. This step-by-step gradual process of fostering independence should also be applied to giving students hints, clues, examples, partial solutions, and other scaffolds.

Precautions and Possible Pitfalls

The guiding questions must be in the teacher's own words and be a direct response to the students' reactions. Care must be taken not to use prefabricated questions that are independent of what is going on in the classroom; this would only create more confusion for the students.

Source

Vygotsky, L. (1978). *Mind in society.* Cambridge, MA: Harvard University Press.

STRATEGY 62: Teach students to ask themselves questions about what they already know about a problem or task they are working on.

What the Research Says

Research was conducted comparing the effectiveness of two types of student self-generated questions: (1) questions designed to enhance understanding of connections of ideas within a lesson and (2) questions designed to access students' prior knowledge/experience and to promote understanding of the connections between that prior knowledge and material in the lesson. All students were trained to give explanations. Teachers were trained by the researcher to teach students how to ask these types of questions and how to give explanations. As part of their regular classwork, students learned, practiced, and were tested on their self-questioning. The results showed that although both types of self-questions led students to develop more complex knowledge, the prior knowledge questions enhanced content learning more effectively.

Teaching to the NCTM Standards

The NCTM Problem Solving Standard calls for students to "apply and adapt a variety of appropriate strategies to solve problems."[5] Students who access prior knowledge can choose among successful strategies employed in the solution of similar problems. This, coupled with their willingness to approach problem solving through a variety of perspectives, should greatly enhance their likelihood of success. When delivering classroom instruction, teachers should stress "asking the right question" so that students can model this on their own. Employing the Socratic method or "Teaching by Questioning" embeds in students' minds the importance of taking a step back to analyze a problem and asking themselves some important questions that are key to formulating a solution.

Classroom Applications

Teach students to question themselves about what they already know about a topic and how this knowledge relates to the current problem/situation. For example, "What do I know about this type of problem? How have I solved problems like this before?" Then teach them to ask themselves how this information applies to the current problem or situation. For example, "How can I use that approach in this situation?"

Student-generated self-questions are very effective thinking and learning tools. These self-questions need to be formulated for the specific problem or task the student is working on. Student-generated questions are superior to teacher-generated questions that are given to students to use. Not all self-questions are equally valuable. When doing problem solving in a mathematics course, there are many times students should ask themselves questions. This form of silent questioning is an excellent way of guiding oneself through the solution of a problem. This sort of self-questioning replaces the typical teacher questioning and begins to develop a problem-solving independence in the student. Not only do crazy people talk to themselves, people solving a mathematics problem do so as well!

Precautions and Possible Pitfalls

Questions generated by the students themselves are more effective than questions provided to them by the teacher, and although student questions are often unpolished and may even sound inaccurate, they understand them and often resent having them sharpened by the teacher.

Source

King, A. (1994). Guiding knowledge construction in the classroom: Effects of teaching children how to question and how to explain. *American Educational Research Journal, 31*(2), 338-368.

STRATEGY 63: Emphasize the general principles that underlie solving specific types of problems.

What the Research Says

Many studies have shown that there are important differences between experts and novices in how they solve problems. One important difference is that experts think about what principle could be applied to solve a problem, whereas novices think about the superficial aspects of a problem.

In one study, forty-five undergraduate physics students were classified into three groups according to the reasons they gave when asked to compare thirty-two sets of two problems. For each set of problems they had to decide whether they would be solved similarly. The "surface-feature" group consisted of students who used superficial aspects of the problems as the basis of their reasoning for at least seventeen of the thirty-two sets of problems. The "principle" group consisted of students who used general principles as the basis of their reasoning for at least seventeen of the thirty-two sets of problems. The "mixture" group consisted of students who used a wide range of reasoning strategies, but none was used a majority of the time.

Next, students were given a problem-solving task consisting of four problems. This task also involved a mathematical proficiency task. The results showed that the "surface-feature" group had the lowest level of performance, with a mean of 14% correct. The "mixture" group's mean was 32% correct, and the "principle" group had the highest level of performance, with a mean of 57% correct. These findings suggest that principle use is related to success in problem solving. Other studies have demonstrated that novices can be taught to think and solve problems the same way experts do.

Teaching to the NCTM Standards

The NCTM Problem Solving Standard calls for all students to

- Apply and adapt a variety of appropriate strategies to solve problems;
- Monitor and reflect on the process of mathematical problem solving.[6]

As students prepare to solve specific problem types, having a deep conceptual understanding of the topic coupled with the ability to use a variety of problem-solving techniques will enable them to craft solutions to a particular problem and related problems. The example given below, which employs a repeated application of "working backwards," illustrates how facility with a problem-solving technique or a combination of techniques allows students to solve a wide range of problems that arise in mathematics.

Classroom Applications

Making students aware of strategies, or principles, relevant to problem solving is one of the goals of the Standards of the NCTM. This can best be seen when observing students doing proofs in geometry. Typically, students observe the model proofs presented by the teacher and then try to copy the example. Often this is done with little or no thinking on the part of the student. Students know that proving triangles congruent or similar tends to lead to the desired result and, therefore, they simply do that with the hope that the next step will be the desired conclusion. Often, this is submitted as something that appears senseless. This situation is an excellent opportunity to reinforce the strategy of "working backwards." Have the student go to the desired result (that which is to be proved) and then ask how it can be proved. That is, if one supposes that the desired result is to prove a pair of lines parallel, then one would seek to prove a pair of corresponding angles congruent or a pair of alternate-interior angles congruent, or some other such relationship that can establish parallelism. Then, taking another step backwards, one asks how to get this pair of angles congruent; perhaps by proving a pair of triangles congruent. One then seeks to prove the triangles congruent and must select the proper method. And so it continues in a backwards-reasoning way. Once the reverse path has been established, the proof is written in the forward order and the proof is complete. This strategy, and especially an awareness of its existence, will also go a long way to improving students' understanding of proofs and their proof-writing ability.

Precautions and Possible Pitfalls

One must be aware that not all strategies can be successfully applied to all problem situations. Sometimes more than one strategy can be used with equal facility. Other times, only one may be usable. The more strategies to which students are exposed, the more powerful their problem-solving skills will become.

Sources

Eylon, B., & Reif, F. (1984). Effects of knowledge organization on task performance. *Cognition and Instruction, 1,* 5–44.

Heller, J., & Reif, F. (1984). Prescribing effective human problem solving processes: Problem description in physics. *Cognition and Instruction, 1,* 177–216.

Thibodeau Hardiman, P., Dufresne, R., & Mestre, J. (1989). The relation between problem categorization and problem solving among experts and novices. *Memory and Cognition, 17*(5), 627–638.

STRATEGY 64: Examine your students' knowledge of mathematics and use this information to write challenging word problems that they will enjoy solving.

What the Research Says

By using students' knowledge of mathematics, teachers can develop word problems that are meaningful to students, are challenging to students, and help them achieve at very high levels. Teachers' knowledge about what problems students can solve has a significant correlation with student achievement. Teachers who have such knowledge teach mathematics differently from teachers who do not have such knowledge. A four-year case study was conducted on an elementary school teacher's use of her investigations into her students' mathematical thinking.

Instead of viewing her students as "blank slates," the teacher strongly believed in building on students' existing knowledge. She systematically investigated their mathematical thinking by interviewing them, by regularly asking them to explain how they solved problems, and by asking them to compare one problem solution to another. One way she used information about students' knowledge of mathematics was to develop a curriculum that primarily involved word problems. Although she wrote most of them, her students also successfully wrote word problems for themselves and each other to solve. They wrote different types of problems with varying levels of difficulty. Students were required to be able to solve all the problems they wrote. The teacher used students' activities and interests as the context of the problems she wrote. The results showed that students enjoyed writing problems and wanted to solve challenging problems with big numbers.

Teaching to the NCTM Standards

The NCTM Teaching Principle states that

> Effective mathematics teaching requires understanding what students know and need to learn and then challenging and supporting them to learn it well. . . . Thus, students' understanding of mathematics, their ability to use it to solve problems, and their confidence in, and disposition toward, mathematics are all shaped by the teaching they encounter in school.[7]

When students are given problems that are routine or repetitive, they quickly lose interest as they are keenly aware that they are not acquiring new knowledge. However, when students are presented with problems that incorporate their interests, challenge them, and extend their knowledge, they value the experience. Having students engage in problem creation is another activity that furthers understanding and comprehension. Problems that are generated, articulated, and solved by students themselves address many of the NCTM Principles and Standards, including the Problem Solving Standard, the Communication Standard, the Learning Principle, and the Teaching Principle.

Classroom Applications

As compared with several decades ago, today, with the advent of the calculator and computer, the teacher has the freedom to use real-world problems with "ugly numbers" that do not calculate to nicely rounded-off numbers, as was the case before such nifty calculating devices were readily available to all students. The question then is, where can the teacher find the sources for problems to which the students can easily relate? For starters, the teacher should get to know her or his students as people. Find out what they like to do in their spare time, for example, what television shows they watch, what comics they read, what movies and video games they like. Building such interests into word problems can make them fun for students to solve. The teacher can also keep up with the local newspapers for the sports events that seem to interest the youngsters, or any unusual sales being run by local merchants. The larger area (or national) newspapers can be consulted, for on practically every page there could be interesting ideas for mathematics applications. Sometimes the teacher will have to add some imagination or take the situation described one step farther to speculate, "What might happen if . . ." In any case, there is a wealth of information available for a teacher to discover ideas to create problems. Teachers who do not see this as an activity that comes easily would be wise to form a small committee of teachers to work together on this effort. In a short time, teachers will get the hang of it and be able to continue independently. The results in terms of student interest will be worth the effort expended.

Precautions and Possible Pitfalls

The primary precaution here is to bear in mind that the topics are to be of interest to the students and not necessarily to the teacher. There are times when this mix-up cannot be avoided, but by being conscious of this possible pitfall, you go a long way toward selecting appropriate examples.

Source

Fennema, E., Franke, M. L., Carpenter, T. P., & Carey, D. A. (1993). Using children's mathematical knowledge in instruction. *American Educational Research Journal, 30*(3), 555–583.

STRATEGY 65: Structure teaching of mathematical concepts and skills around problems to be solved, using a problem-centered or problem-based approach to learning.

What the Research Says

Problem-centered or problem-based learning is becoming recognized as an outstanding way of teaching both content and problem-solving skills. One study compared six classes that received problem-centered mathematics instruction for two years with students who received problem-centered mathematics instruction for one year and with students who received traditional textbook-based instruction. Researchers examined students' performance on standardized achievement tests and investigated students' personal goals and beliefs about the reasons for their success in mathematics. The results showed that students who received problem-centered instruction for two years demonstrated significantly higher mathematics achievement than traditionally instructed students, in terms of both their proficiency in solving problems and their conceptual understanding. In addition, problem-centered-learning students had stronger beliefs than traditional students about the importance of finding not only *different* ways of solving problems, but the importance of finding *their own ways* of solving problems. Students who received problem-centered mathematics instruction for only one year and then returned to textbook teaching performed at levels comparable to the textbook-instruction-only students. Consequently, in order to achieve meaningful benefits from the problem-centered approach, students should receive more than one year of instruction using this form of teaching. This problem-centered type of approach has become standard instructional practice in many medical school programs.

Teaching to the NCTM Standards

The NCTM *Principles and Standards for School Mathematics* state that

> Students should view the difficulty of complex mathematical investigations as a worthwhile challenge rather than an excuse to give up. Even when a mathematical task is difficult, it can be engaging and rewarding. When students work hard to solve a difficult problem or to understand a complex idea, they experience a very special feeling of accomplishment, which in turn leads to a willingness to continue and extend their engagement with mathematics.[8]

As the application below suggests, introducing a topic with a provocative problem creates an interest that is motivational and gives purpose to the day's lesson.

Classroom Applications

Instead of starting a unit by using the textbook and telling students about a mathematical topic and explaining and demonstrating various concepts, problems, and solution methods, start by giving students a meaningful problem to solve. Problem-centered or -based learning is a method of teaching that involves using ill-structured, real-world problems as the context for learning basic content through in-depth investigations. In order to solve the problem, students will need to learn specific mathematical concepts and solution strategies. Teachers give students only enough information to enable them to begin their inquiry. They never give students enough information to actually solve a problem. With this instructional method, students cannot simply solve problems by applying a particular formula. There must be student reasoning and inquiry. Often there is more than one way to solve a problem. Students will learn important concepts and skills through mathematical inquiry in a meaningful context. The teacher's role is to be a coach, mentor, or tutor who guides students in their inquiry and helps them develop and understand their own thinking.

One topic that lends itself to this sort of problem-solving investigation is the consideration of maxima and minima. For example, students could seek to maximize an area with a given perimeter. They could consider the shape as a variable. They may have it as a rectangle or a polygon. They may even consider a circle with this given perimeter to see how it compares to the other shapes. This topic can be considered before the study of calculus by inspecting the turning point of a parabola, or by merely inspecting extremes to see the behavior of the variable.

Precautions and Possible Pitfalls

⚠ Problems must be at an appropriate level of complexity. In addition, students must have appropriate prior knowledge so they know or can figure out what they need to learn in order to solve the problem. The topic selected must be appropriate for the ability level of the students, and above all it must garner the proper interest among the students. Without this it will not serve the desired goals, mentioned above.

Sources

Checkley, K. (1997, Summer). Problem-based learning: The search for solutions to life's messy problems. *Curriculum Update: Association for Supervision and Curriculum Development.*

Lambros, A. (2004). *Problem-based learning in middle and high school classrooms: A teacher's guide to implementation.* Thousand Oaks, CA: Corwin.

Wood, T., & Sellers, P. (1996). Assessment of a problem-centered mathematics program: Don't always call on volunteer responses to teacher questions. Third grade. *Journal of Research in Mathematics Education, 27*(3), 337–353.

Wood, T., & Sellers, P. (1997). Deepening the analysis: Longitudinal assessment, a problem-centered mathematics program. *Journal of Research in Mathematics Education, 28*(2).

STRATEGY 66: Help students learn without relying on teacher-centered approaches. Give them carefully chosen sequences of worked-out examples and problems to solve.

What the Research Says

Students learned to solve problems from working with examples and from learning by actually doing them. Students understood what procedures were needed (as revealed by their explanations) and knew how to apply them; they were not merely memorizing procedures. Middle school students (N = 118, age = 13 years) were studied when solving problems involving simplifying fractions, factoring quadratic expressions, and manipulating terms with exponents, and when solving geometry problems. Twenty students were studied by asking them to think out loud while working individually. Students were divided into two groups. One group was in the learning-by-doing condition; the other group was in the learning-from-examples condition. Ninety-eight students were studied in their regular classroom settings using the learning-from-examples method. Researchers examined the levels and speeds of learning in the classroom.

The researchers concluded that working with examples and learning by doing are effective alternatives to teacher-centered instructional approaches such as lecturing or other methods of direct instruction.

Teaching to the NCTM Standards

The NCTM Problem Solving Standard requires that students "solve problems that arise in mathematics and in other contexts and apply and adapt a variety of appropriate strategies to solve problems."[9] The NCTM recommends that problem solving be taught by "posing questions and tasks that elicit, engage, and challenge each student's thinking."[10] This cannot be done using a lecture model; it is much more effectively taught by employing group work and discussions that compel students to justify their ideas. It is the role of the teacher to decide how to redirect responses, synthesize student contributions, and allow the class to reason together to successfully solve the problem at hand:

> Decisions about when to let students struggle to make sense of an idea or a problem without direct teacher input, when to ask leading questions, when to tell students something directly are crucial to orchestrating productive mathematical discourse in the classroom.[11]

Classroom Applications

Perhaps a good illustration of where the lecture method works worst is in the instruction of problem solving. Although problem solving has continuously been stressed for importance by the NCTM since before it issued its Agenda for Action in 1980 and later, of course, in its Standards in 1989, it is often a neglected topic in the curriculum. Problem solving is a very individual activity. Everyone learns at his or her own pace and with a different experiential background. Therefore, it is not advisable for the teacher to teach this topic through lecture. By providing students with model solutions, after they have had time to try to solve (or actually solve) a problem, you enable the students to focus on the various parts and skills of the solution on their own terms and at their own pace. This allows each student to enjoy the solution and then to ponder over its cleverness, with the hope that this will allow a deeper appreciation for the work and for replication in the near future.

Like musicians hoping to play at Carnegie Hall, to be good problem solvers students need practice, practice, and more practice! Learning to solve problems by doing the problems rather than watching and listening to the teacher is an important ingredient of success.

Precautions and Possible Pitfalls

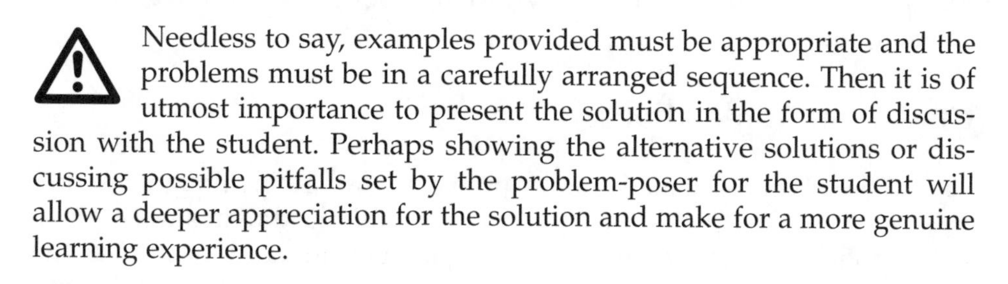

Needless to say, examples provided must be appropriate and the problems must be in a carefully arranged sequence. Then it is of utmost importance to present the solution in the form of discussion with the student. Perhaps showing the alternative solutions or discussing possible pitfalls set by the problem-poser for the student will allow a deeper appreciation for the solution and make for a more genuine learning experience.

Source

Zhu, X., & Simon, H. A. (1987). Learning mathematics from examples and by doing. *Cognition and Instruction, 4*(3), 137–166.

STRATEGY 67: Students need time to practice planning their solutions to problems.

What the Research Says

Students often have difficulty solving problems because they start working on a solution as soon as they finish reading a problem. Instead, they should try to understand the problem and then develop a problem-solving plan, like experts do. Students need to be taught the importance of planning the solving of problems and be provided with time to practice planning their solution attempts. One study involved teaching a group of students to plan their problem solving in preparation for an exam they were going to take. Another group of students studied as they normally did for the exam. Students who were in the planning group performed better than those who used their traditional study methods, even though this group of students reported spending more time studying than did students in the planning group.

Teaching to the NCTM Standards

The NCTM Learning Principle stresses the importance of "conceptual understanding in becoming proficient in a subject."[12] Students must function beyond the procedural level or they will not attain the "flexible knowledge" that can be applied to a variety of situations. In order to increase conceptual understanding, teachers are encouraged to have students gain proficiency in the planning of a solution to a problem. Teachers should have a classroom environment where a variety of approaches to problem solving are considered, compared, contrasted, and valued. Students will soon realize that most problems can be solved many different ways, though some are more elegant and efficient than others.

Classroom Applications

The solution of a mathematics problem requires forethought, as indicated above. One of the best ways to instill this trait in students is to model it in classroom activities. When tackling a problem with the class, rather than jumping right into solving it, the teacher should think out loud, allowing the students to follow the teacher's thought process on how he or she will solve the problem. Or, the teacher can conduct a discussion about how to go about finding a path to the solution. Students should see that planning must precede problem solving. Students also need a repertoire or "tool chest" of problem-solving strategies to apply to problems they are going to solve. Such strategies might include working backwards, finding a pattern, adopting a different point of view, solving a simple analogous problem, using extreme cases, visually representing a problem, intelligent guessing, estimating and testing, accounting for all possibilities, organizing data, and deductive reasoning. For a detailed discussion of these problem-solving strategies, we recommend Posamentier and Krulik (1996).

In addition to planning such problem-solving strategies, other good planning techniques for students to use when solving mathematics problems are as follows:

1. Ask students to represent (in pictures or symbols) and describe the problems they are solving (including what the relevant information is, what is irrelevant, and what type of problem it is) and how they solve problems.
2. Have students compare their solution methods to those of an expert.
3. Have students compare how they solve one problem with how they solve other problems, some of which are similar and some of which are different.
4. Instead of having students carry out a solution, have them select a sequence of the operations needed to solve a problem. Have them refer back to the problem statement to make sure that the plan fits the problem.

Make sure students understand when, why, and where to apply mathematical procedures, concepts, algorithms, and strategies. Have students keep journals where they describe and evaluate their understanding of concepts and how they plan, monitor, and evaluate their problem solving. Journals should include descriptive and evaluative information. Descriptive information includes students' explanations of what their approach is, when, where, and why they used it for the particular problem, and how they used it. Evaluative information includes whether the approach was effective, if there is anything they would do differently next time, and if there are any other ways they could have solved the problem. Periodically collect their journals and give them feedback.

Precautions and Possible Pitfalls

When demonstrating how to plan problem-solving approaches, be sensitive to students' different learning rates. Avoid rushing! Don't let the pace of the faster students push you to a speed that would be hurtful to slower students. If time is taken for proper modeling, the slower students can learn to master what you are teaching them about how to plan problem solving, and they might even learn this with a clearer understanding than the faster students. A rushed time factor could be quite detrimental to all the students' learning patterns. Do not group problems into sets where all the problems are solved the same way. This can create a problem-solving set, promote planning solutions by rote, and discourage thinking.

Sources

Nickerson, R. S., Perkins, D. N., & Smith, E. E. (1985). *The teaching of thinking.* Hillsdale, NJ: Lawrence Erlbaum.

Posamentier, A. S., & Krulik, S. (1996). *Teacher! Prepare your students for the mathematics for SAT* I: Methods and problem-solving strategies.* Thousand Oaks, CA: Corwin.

6

Considering Social Aspects in Teaching Mathematics

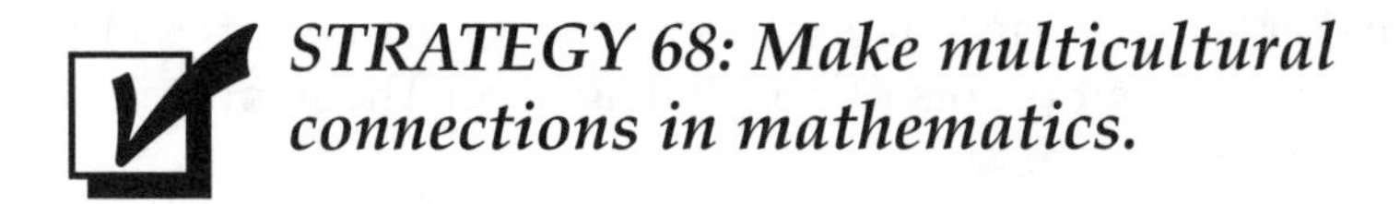

STRATEGY 68: Make multicultural connections in mathematics.

What the Research Says

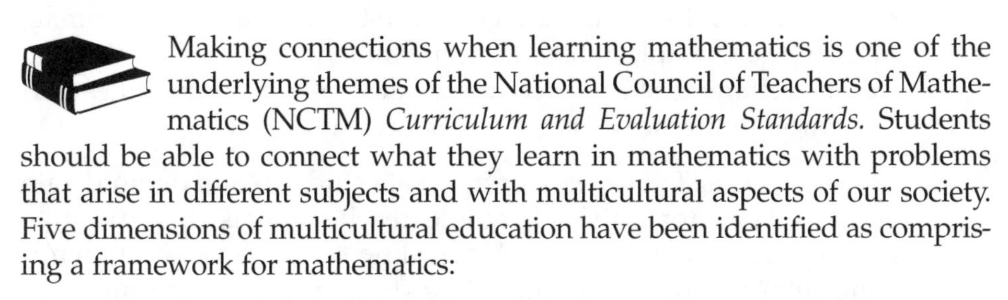

Making connections when learning mathematics is one of the underlying themes of the National Council of Teachers of Mathematics (NCTM) *Curriculum and Evaluation Standards.* Students should be able to connect what they learn in mathematics with problems that arise in different subjects and with multicultural aspects of our society. Five dimensions of multicultural education have been identified as comprising a framework for mathematics:

- *Integrate content* reflecting diversity when teaching key points.
- *Construct knowledge* so students understand how people's points of view within a discipline influence the conclusions they reach in a discipline.
- *Reduce prejudice* so students develop positive attitudes toward different groups of people.
- *Use instructional techniques* that will promote achievement from diverse groups of students.
- *Modify the school culture* to ensure that people from diverse groups are empowered and have educational equality.

Teaching to the NCTM Standards

The NCTM *Handbook of Research on Mathematics Teaching and Learning* recognizes that "much mathematical knowledge is acquired outside school."[1] The term *ethnomathematics* refers to "forms of mathematics that vary as a consequence of being embedded in cultural activities whose purpose is other than 'doing mathematics.'"[2]

Students are very much aware of social and cultural issues in a school community. Teachers should be sensitive to this and take steps to build a diverse community of learners. The applications below provide ways to recognize cultural differences and build on them to enhance the teaching of mathematics and make it more meaningful to all of the students in the class.

Classroom Applications

Examples of how to apply the five multicultural dimensions to mathematics are as follows:

1. *Integrate content* so that the history of mathematics includes contributions of mathematicians from many cultures and ethnicities, and is not Eurocentric. Examples: Teach students about Benjamin Banneker, an African American creator of mathematical puzzles; teach them about proofs of the Pythagorean Theorem from China, India, and Babylon.

2. *Construct knowledge* so students see the universal nature of mathematical components of measuring, counting, locating, and designing.

3. *Reduce prejudice* by using mathematics (statistical data) to eliminate stereotypes. Examples: show there are more whites than blacks who are on welfare; teach the economic value of recycling bottles and cans through the story "The Black Snowman."

4. *Use instructional techniques* that motivate students and demonstrate mutual respect for culture. Example: teach students about positive and negative numbers through traveling on a subway and identifying subway stops after and before students board the train. This shows the urban culture and how it can require some mathematical thinking. Similarly, some of the huge skyscrapers have several levels below the street level and there, too, is an urban application for the directed numbers.

5. *Modify the school culture* so students from diverse cultures are grouped together and can participate in extracurricular activities, and so teachers have high expectations for success from all students

regardless of diverse culture backgrounds. Examples: Structure cooperative learning groups heterogeneously using culture and ethnicity as variables; assign African American students as tutors with white and/or Asian students as tutees.

Precautions and Possible Pitfalls

Make sure that multicultural aspects of lessons are not done in a patronizing manner.

Sources

Banks, J. A. (1994). Transforming the mainstream curriculum. *Educational Leadership, 51*(8), 4–8.
Bishop, A. (1988). Mathematics education in its cultural context. *Educational Studies in Mathematics, 19,* 179–191.
Mendez, P. (1989). *The black snowman.* New York: Scholastic.
Moses, R., Kamii, M., Swap, S., & Howard, J. (1989). The algebra project: Organizing in the spirit of Ella. *Harvard Educational Review, 59*(4), 423–443.
Strutchens, M. (1995). *Multicultural mathematics: A more inclusive mathematics.* ERIC Digest. Clearinghouse for Science, Mathematics and Environmental Education, Columbus, Ohio. (ERIC Document Reproduction Service No. 380–295)

STRATEGY 69: Find out about your students' families and how their values and practices might affect students' attitudes and performance in mathematics.

What the Research Says

Similarities and differences were found when comparing families from different cultural backgrounds. Regardless of cultural background, most parents valued schooling, believed their children liked school, and expected their children to get a degree from a four-year college. Families from different cultures differed in their parenting practices, their perceptions of their children's abilities, how they explained students' successes and failures in school, how they rewarded children for their school performance, and in how critical they were of their children's academic performance. Family background also influenced whether children were provided with desks for studying, and whether children studied on Saturdays and Sundays. These findings were from a study of forty-one families from Anglo, Hispanic, and Vietnamese backgrounds. Parents and children were interviewed in their primary language; parents were interviewed separately from their children. The researchers were more struck by the similarities of families, regardless of cultural background, than by the differences between them.

Teaching to the NCTM Standards

The NCTM *Handbook of Research on Mathematics Teaching and Learning* addresses the role of culture in learning mathematics. The study revealed two schools of thought regarding culture and mathematics. One view, which was articulated by Stigler and Baranes (1988), contends that

> Mathematics is not a universal, formal domain of knowledge . . . but rather an assemblage of culturally constructed symbolic representations and procedures for manipulating these representations. . . . As children develop, they incorporate representations and procedures into their cognitive systems, a process that occurs in the context of socially constructed activities. Mathematical skills . . . are forged out of a combination of previously acquired (or inherited) knowledge and skills, and new cultural input.[4]

Another viewpoint, by Gal'perin and Georgiev (1969) and D'Ambrosio (1986), suggests that "the analysis of cultural influences on mathematical knowledge can demonstrate both differences and invariance in mathematical knowledge across cultures."[5] Even within cultures, the practice of mathematics can vary according to its purpose. One of the invariants in the cultural analysis of mathematics was the commonality of its importance. Because all cultures value the teaching and learning of mathematics, it is advisable for teachers to communicate with families to reinforce the partnership that should exist between the school, the home, the teacher, and parents.

Classroom Applications

There are clearly cultural differences with regard to the emphasis on topics and skills taught in school mathematics. Whether these cultural differences are related to youngsters by their parents is something the teacher would be wise to discover through interviews with individual students and in teacher-parent conferences. There may be an unduly high priority on arithmetic skills stressed in one home, while in another this skill may be somewhat neglected in favor of an understanding of concepts such as proportional thinking. This latter concept is far more prominent in the school curriculum of central European schools than it is in the United States. Furthermore, the time available for study can influence the achievement of a youngster. If the teacher finds that the children have more time available for homework on the weekend, then he or she may use this opportunity to assign more work during that time frame, in addition to maintaining regular homework assignments during the week. If this is not the case, because of religious reasons, for example, the student must pay the price for this misjudgment. In short, teachers should communicate with the home, via the students and the parents, so that the classroom support at home is used properly.

Precautions and Possible Pitfalls

Be careful not to take at face value comments made by students about their home life interfering with the school's homework assignments. Students, who see the teacher as very sensitive to being obtrusive to the student's home life, may intentionally try to block progress with "harmless untruths." Precaution is appropriate here!

Sources

D'Ambrosio, V. (1986). Socio-cultural bases for mathematics education. In M. Carss (Ed.), *Proceedings of the Fifth International Congress of Mathematical Education* (pp. 5–9). Boston: Birkhauser.

Gal'perin, P., & Georgiev, L. (1969). The formation of elementary mathematical notions. In J. Kilpatrick & I. Wirzup (Eds.), *The learning of mathematical concepts. Soviet studies in the psychology of learning and teaching mathematics* (Vol. 1). Palo Alto, CA: SMSG.

Lindholm, K. J., & Miura, I. T. (1991). *Family influences on mathematics achievement: Anglo, Hispanic and Vietnamese students.* Paper presented at the annual meeting of the American Psychological Association, San Francisco.

STRATEGY 70: Reach out to parents to form a partnership for educating elementary and high school students.

What the Research Says

Students want their parents to be involved in their education. A high level of parental involvement in children's education generally leads to a high level of academic achievement. Parents frequently are involved with their children's education while children are in elementary school, but often stop being involved once children are in high school. One study looked at 748 urban elementary and secondary school students' (Grade 5, $N = 257$; Grade 7, $N = 257$; Grade 9, $N = 144$; and

Grade 11, $N = 90$) requests for and attitudes about their families' involvement in their education. Of these, 449 were black, 129 were Hispanic, and 121 were white. The study compared high- and low-achieving students in mathematics and English (reading for elementary school students). It also examined whether there were ethnic differences in students' feelings about family involvement. Students in all grades requested parental assistance with schoolwork and had positive attitudes about using their parents as educational resources, although elementary students made more requests and had more positive attitudes than secondary school students. Both high- and low-achieving students showed interest in parental involvement. However, at the elementary school level, high-achieving Hispanic students in mathematics had more favorable attitudes than did low-achieving Hispanic students in mathematics. Black and Hispanic students were generally more interested in parental involvement than were white students.

Teaching to the NCTM Standards

The NCTM *Handbook of Research of Mathematics Teaching and Learning* examines parental and societal influences on "mathematics attitudes and performance."[6] The *Handbook* cites the study of Eccles and Jacobs (1986) that "identified parents as a critical force"[7] in a child's mathematics education. They stated, "Parents exert a more powerful and more direct effect than teachers on children's attitudes toward mathematics . . . parents' gender-stereotyped beliefs are a key cause of sex differences in students' attitudes toward mathematics."[8] The findings of Stevenson and Newman (1986) "indicated that not only parents affect their children's long-term achievement and attitudes in mathematics but also that the patterns identified differed for mathematics and reading."[9] Other studies "found students' attitudes towards mathematics and their decisions to continue with mathematics were linked to their parents' conceptions of the educational goals of school mathematics and the perceived relevance to their children's long-term life goals."[10] Because all of the research supports a correlation between parental attitudes and success in mathematics, reaching out to parents is a crucial step in forming a partnership that can help shape positive attitudes and student success in mathematics.

Classroom Applications

Teachers should reach out to parents to enhance their involvement and develop a partnership for their children's education. Many parents are unaware that they have the ability to have an impact on their children's education even if they are not well educated themselves. Teachers can explain and illustrate for parents how a parent can function as an educational manager and/or teacher. Some examples of the parent as manager follow:

- Provide a time, quiet place, and adequate light for studying. Help the child determine the best time and place to work.
- Each night ask if there is a homework assignment and ask to see it when it has been completed.
- Each night ask about what happened in school.
- Have a dictionary accessible and encourage the child to use it.
- Find out when tests are to be given, and make sure the child has a good night's sleep and breakfast the day of the test.
- Visit the school to discuss the child's progress and to find out what can be done at home.
- Communicate positive attitudes and expectations about the child's school performance.
- Avoid letting a child's household responsibilities assume more importance than schoolwork.

Teachers can prepare a handout for parents so they have some idea about what they can do at home to support this partnership.

Precautions and Possible Pitfalls

⚠ If parents do not speak English well, they may be reluctant to communicate with teachers. In such cases, if the teacher cannot speak the parents' language, a community volunteer might act as a school advocate and resource, or someone from the school might be able to translate a letter or handout for parents into the parents' native language.

Sources

Hartman-Haas, H. J. (1983). *Family educational interaction: Focus on the child.* Paper presented at the annual meeting of the American Educational Research Association, Montreal, Canada.

Hartman-Haas, H. J. (1984). Family involvement tips for teachers. *Division of Research Evaluation and Testing Research Bulletin* [Newark Board of Education, Newark, NJ], 1–12.

Jacobs, J. E., & Eccles, J. S. (1985). Gender differences in ability: The impact of media reports on parents. *Educational Researcher, 14*(3), 20–25.

Stevenson, H. W., & Newman, R. S. (1986). Long-term prediction of achievement in mathematics and reading. *Child Development, 57,* 646-659.

STRATEGY 71: Inform parents that they should not let media reports about studies of other children change their views of their own children's abilities to be successful in mathematics.

What the Research Says

Parents may develop misconceptions about their children's abilities as a result of reports in the media. One study examined the impact on parents of a media report on gifted junior high school students. Extensive media coverage focused on a report of a major gender difference in students' mathematical aptitudes. The study compared parents' views about their children's mathematical aptitudes before and after exposure to the media report. The results showed that the media coverage changed parents' attitudes about their children's mathematical abilities. Fathers of sons and mothers of daughters developed stronger sex-based stereotyped beliefs after the media coverage.

Teaching to the NCTM Standards

The media play an important role in shaping our perceptions regarding almost all aspects of our lives. The success of our educational system is not exempt from public scrutiny and media attention. Recently, alarming conclusions about mathematics education in the United States has emerged from international studies that compare our system with those of other developed nations. The Third International Mathematics and Science Study (TIMSS) assessed the mathematics and science performance of U.S. students in comparison to their peers in other nations. The results were widely reported in the media as the United States scored markedly lower than Korea, Singapore, and Japan. There was a call to action to investigate and identify why the United States lagged behind other industrialized nations and what reforms could be undertaken to narrow the achievement gap. While self-evaluation is always a healthy component of curriculum implementation and revision, institutions, teachers, parents, and students should be cautioned about overreacting to every story that is reported by the media. Internalizing these reports can have a demoralizing effect on mathematics education. If parents lower their expectations for mathematics instruction for their children and have little confidence in the curriculum that is in place at their school, they ultimately lower expectations for their own children. This can seriously impact the effectiveness of mathematics instruction. Parents are advised to react cautiously to reports of both successes and failures in a system. Success stories are often obscured by reports of isolated instances of failure, and failures are obscured by media reports of "successful" math programs. The NCTM supports obtaining input from many sources, such as teachers, math educators, mathematicians, educational researchers, and the general public. It is the thoughtful analysis and synthesis of reports from a variety of sources that should shape our attitudes and perceptions about the effectiveness of mathematics instruction. Of course, one of the best ways for parents to feel secure about their child's mathematics instruction is to visit the class and draw inferences from the classroom environment and the instructional model that the teacher is using. Teachers should invite parents to observe the class during Open School Week and hold workshops with parents to strengthen the parent-teacher partnership.

Classroom Applications

For most people, mathematics is a source of frustration. A common reaction to mathematics by many parents is to say that they did poorly in the subject themselves, so they tend to accept this from their children. It appears to be a sort of "badge of honor" to admit weakness in mathematics (unlike almost any other subject). Teachers would be wise to offer periodic workshops for parents, keeping them informed of what is being taught, how it is being presented, and what can be expected of their children, both in performance and results. This sort of workshop experience will also give teachers an opportunity to communicate with parents regularly and to inform them of their individual child's progress and ability to be successful in mathematics. Then parents will be more prepared to interpret reports from the media and other sources. They would also be less likely to succumb to overgeneralizations and stereotypes that could undermine their child's performance in mathematics.

Precautions and Possible Pitfalls

Extreme patience must be used when working with parents. Recognize that many of them may have been away from a school setting and the concomitant behavior for many years. Be cautious when reporting frequently on a student's progress; leave room for improvement and never "close the door" on an individual student, no matter how frustrating the child's progress may be. Remember that some parents have a tendency to overreact to the teacher's comments, and that may have deleterious effects.

Source

Jacobs, J. E., & Eccles, J. S. (1985). Gender differences in ability: The impact of media reports on parents. *Educational Researcher, 14*(3), 20–25.

STRATEGY 72: Some students do not think they have control over their academic successes and failures. Help these students recognize that they do have some control.

What the Research Says

Research has demonstrated that students differ in their perceptions of control over their successes and failures. Some students externalize responsibility (externalizers) while other students internalize this responsibility (internalizers). Externalizers assume powerful other people, such as teachers and parents, are responsible for their performance. Externalizers often also assume that chance plays an important role in their destiny. Internalizers perceive that they have control over their own destiny and are responsible for their own performance.

A study was conducted with 198 male and female college students; some were externalizers and some were internalizers. Before hearing a lecture, a group of internalizers and externalizers was shown an eight-minute color videotape of a psychology professor who told about his freshman year in college. He kept failing and persisted only because a friend urged him to. Eventually he succeeded as an undergraduate and as a graduate student. He encouraged students to attribute poor performance to not making enough effort. He also encouraged them to attribute good performance to making appropriate effort and to ability. He explained that students can change the amount of effort they devote to a task and that a major component of successful effort is persistence. Finally, he emphasized that long-term effort improves ability.

Another group of internalizers and externalizers heard the same lecture, but they did not see the videotape. After the lecture all students were given homework unrelated to the lecture and were told that they would be tested on the homework and the lecture in a week. The results showed that the videotape improved the performance of externalizers on both the homework and lecture tests. Both internalizers and externalizers who saw the videotape performed better on the homework and lecture tests than students who did not see the videotape.

Teaching to the NCTM Standards

The NCTM Assessment Principle mandates that "assessment should support the learning of important mathematics and furnish useful information to both teachers and students."[11] This formative assessment should shape the teaching and pedagogical practice of the classroom teacher. Students who have difficulty on traditional assessments may have valuable insights and understanding of the mathematics that is not evident through traditional assessments. While constructed, open-ended responses allow for more insight into student comprehension, students can be assessed informally through teacher questioning, observations, careful (teacher) review of homework, group work, math projects, and journal writing. The goal of assessment is to determine what students know; it should not focus solely on what they do not know. Mathematical success should empower students to strive for success, measured by a variety of assessment instruments. Students will grow if they are shown that their hard work and efforts are recognized by the classroom teacher, who aggressively looks for evidence of learning instead of approaching assessment through the narrow viewpoint of a multiple-choice, short answer, recall, or speed drill examination.

Classroom Applications

Specifically acknowledge when students make sincere efforts and when they use effective and/or ineffective strategies. It is often difficult for students to be aware of their physical growth or, for that matter, of their learning achievement (or growth). Every once in a while, it would be wise for a teacher to demonstrate this growth to students. Feedback on growth should include progress in understanding, remembering, and applying specific concepts, strategies, and skills. There are various ways that this can be done. Give a pre-test and post-test. Videotape students working at the beginning, middle, and end of a unit of study. In both of these cases students can see their achievement for themselves and that they themselves were largely responsible for the progress. Take for example the solution of a quadratic inequality. At first sight, it would clearly be confusing, and this confusion, often frustrating, should be gently documented. If possible, this process of dissecting the quadratic into factors, inspecting their relationship to the original quadratic, graphing the result, and through several exercises showing a clear ability to solve quadratic inequalities, gives the teacher ample ammunition to convince students that through their efforts achievement is possible and that they control it.

Precautions and Possible Pitfalls

Teachers should take care to make these external "interferences" (pre-test, post-test, videotaping, etc.) as unobtrusive as possible so that they document normal behaviors. Furthermore, the teacher should constantly bear in mind the goal of the assessment, that it is not a rating device, but rather one that serves a very specific goal—namely, demonstrating to students that genuine learning has taken place and, for the most part, they are responsible for that growth.

Some students develop feelings of hopelessness or helplessness from repeated poor performance in mathematics. These students tend to doubt that they have the ability to succeed in mathematics. These students, in particular, need special attention in developing the feeling that they have the power to improve their performance by making persistent efforts and by using more effective strategies.

Source

Perry, R. P., & Penner, K. S. (1990). Enhancing academic achievement in college students through attributional retraining and instruction. *Journal of Educational Psychology, 82*(2), 262–271.

STRATEGY 73: Teach students, especially girls, to believe that success in mathematics results from their efforts.

What the Research Says

Recent research shows that one of the major factors influencing students' motivation is how they explain their successes and failures. There are four explanations that students commonly give for their academic successes and failures: ability, effort, task difficulty, and luck. In one study, 279 junior high school students completed a questionnaire on their beliefs about mathematics achievement. The questionnaire was completed both before and after a mathematics exam. The results showed that girls tend to attribute failure in mathematics to low ability and bad luck and they tend to attribute success in mathematics to high ability. Also, girls were more likely than boys to hide their papers after failure and to have less pride in their successes. These results lead to the conclusion that how students explain their successes and failures in mathematics depends upon students' gender. Teachers should help students in general, and girls in particular, to realize that they have the ability to control their own academic destiny, and that the specific efforts they make, or don't make, lead to specific outcomes.

Teaching to the NCTM Standards

The NCTM Equity Principle states, "Excellence in mathematics education requires equity—high expectations and strong support for ALL students."[12] In order to achieve excellence in mathematics for all students, teachers must make a connection with each student by providing an approach that will be most effective for the student's particular learning style and providing feedback that is useful to the student. This "formative assessment" should enable students to perform an error analysis to determine precisely why they obtained the wrong answer. In addition, this feedback should help students find the correct solution by pointing them toward similar problems that have been solved correctly. In many instances, students who have traditionally struggled in mathematics have been "labeled" and are not challenged to meet high standards. Teachers must maintain high expectations for these students and provide encouragement to augment alternative teaching strategies that should be employed. Similarly, gifted students must be provided with enrichment so that they are actively engaged and challenged mathematically.

Classroom Applications

Teachers can teach students how to convert failures into future successes by doing an error analysis using the following model:

- What was the incorrect answer? What is the correct answer?
- Why did you have the wrong answer? Be specific.
- How can you prevent that same type of mistake in the future? What did you learn that you need to remember for future success?

Have students compare an example of their successful performance with an example of their unsuccessful performance so they can determine how the situations differed. Help students identify patterns in their successes and failures so they can repeat the use of successful strategies with confidence and overcome failures with more effort and better strategies. Teachers can generate some positive feelings in the learner by presenting some simple exercises, ones where success is reasonably expected. Hopefully, if presented cleverly, these simple exercises will generate a positive attitude that will carry over to somewhat more complicated problems or exercises on the same topic.

Precautions and Pitfalls

Sometimes students show a pattern of "learned helplessness" and feel they have no control over their own destiny. These students usually feel that other powerful people, such as the teacher, and luck or fate determine what happens to them. These students tend to externalize responsibility for and control over their academic performance. They may also feel that they do not have the ability to succeed. Such students need special guidance in learning how specific thinking, learning, and problem-solving strategies lead to specific academic outcomes. When giving exercises designed to boost positive attitudes, make sure they are not too simple. If success comes too easily, it will not have much significance or impact.

Source

Stipek, D. J., & Gralinski, J. H. (1991). Gender differences in children's achievement-related beliefs and emotional responses to success and failure in mathematics. *Journal of Educational Psychology, 83*(3), 361–371.

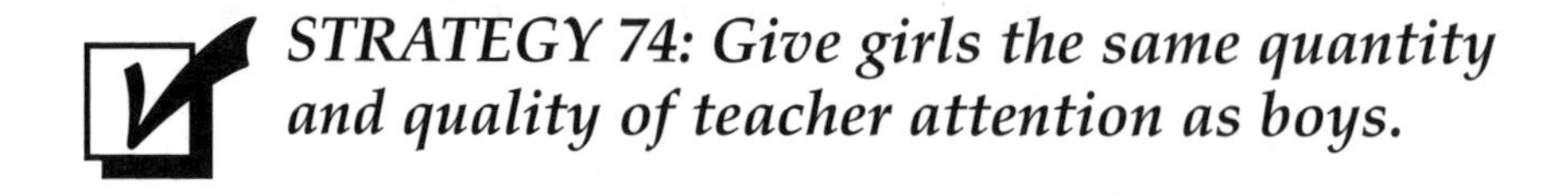

What the Research Says

A study was conducted in ten high school geometry classes in two school systems. One school system was in an urban-suburban community, with well-educated and relatively wealthy people. The other was a lower class, less well-educated rural community. Each of ten mathematics teachers was observed ten times for a total of 100 observations. The researchers used the Brophy Good Teacher-Child Dyadic Interaction System to study teacher-student contacts or interactions. They looked at and compared the quantity and quality of teachers' interactions with girls and boys. The results showed that teachers treated girls and boys differently in several ways, including responding to questions, the cognitive level or difficulty of questions asked, praise and criticism, encouragement, individual help, and conversation and joking. Generally, teachers acted in more positive ways with boys than girls. Boys received more attention, more reinforcement, and more emotional support.

Teaching to the NCTM Standards

The NCTM *Handbook of Research on Mathematics Teaching and Learning* states: "There are a number of ways in which schools, and teachers within the schools, differentiate between groups of male and female students. The former do so through their organizational procedures, the latter through their behaviors, expectations and beliefs."[13] The *Handbook* cited research that

> confirmed that there were differences in the ways teachers behaved, on average, towards the two groups of students. Males, they reported, tended to receive more criticism, be praised more frequently for correct answers, have their work monitored more frequently and have more contacts with their teachers. The sex of the teacher did not seem to affect these conclusions.[14]

The NCTM equity principle states that, "Excellence in mathematics education requires equity—high expectations and strong support for all students."[15] The application below suggests that teachers must make conscious efforts to ensure that female and male students are treated equally.

Classroom Applications

Teachers should make every effort to consciously and conscientiously encourage girls during regular classroom instruction. Boys should not be allowed to select the best seats in the classroom or to dominate students' responses. Teachers should take care not to show gender bias and should attempt to identify famous female mathematicians from the past and highlight their contributions to the development of mathematics. There are a number of books available from professional organizations and commercial publishers that address the issue of women in mathematics. For example, from the National Council of Teachers of Mathematics (1906 Association Drive, Reston, VA 20191-1593) the following titles are available: *Celebrating Women in Mathematics and Science, Multicultural and Gender Equity in the Mathematics Classroom, New Directions for Equity in Mathematics Education,* and *Reaching All Students With Mathematics.*

Precautions and Possible Pitfalls

Teachers should make this inclusion of gender issues a natural part of the instructional program and not merely an add-on. The latter would seem obvious and unnatural and would defeat the purpose. Cultural influences often negatively affect girls' attitudes about and their performances in mathematics.

Source

Becker, J. R. (1981). Differential treatment of females and males in mathematics classes. *Journal for Research in Mathematics Education, 12*(1), 40–53.

STRATEGY 75: Make special efforts to encourage girls to study mathematics.

What the Research Says

A study was conducted with 400 students in nine high school algebra and geometry classes with ninth, tenth, and eleventh graders. One group participated in a special workshop designed to increase girls' appreciation for and interest in mathematics. The workshop was based on the videotape, "Multiplying the Options and Subtracting the Bias." After viewing the videotape, students participated in a student-centered discussion. The other group did not participate in a special workshop. All students completed a questionnaire on their plans for high school and after high school, before the workshop and four weeks after the workshop. The results showed that girls who participated in the workshop indicated:

- They were going to study more mathematics both during and after high school.
- They actually did enroll in more mathematics courses in high school.
- They changed in their perception of the usefulness of mathematics for their future lives, and this perception was significantly related to changes in their plans to study mathematics.

Teaching to the NCTM Standards

The NCTM *Handbook of Research on Teaching and Learning* highlighted the link between confidence and achievement in mathematics. The research "revealed that males in grades 6 to 12 consistently showed greater confidence in their ability to do mathematics than did their female classmates,"[16] and "males' and females' differing attitudes towards mathematics were paralleled by differences in the test performance of the two groups."[17] It was summarized by Joffe and Foxman (1988):

> It seems reasonable to assume that some aspects of attitude, perhaps confidence, may be operating—positively in the case of boys, negatively in the case of girls. It also seems likely that if these feelings are already in operation at age 11 that they will become much stronger by age 15 unless some intervention is made.[18]

It is extremely important for teachers to take proactive measures to give female students leadership roles in the classroom. Relegating female students to secretarial tasks in the classroom only reinforces the institutional stereotypes that prevent women from attaining the self-confidence that correlates to success in higher mathematics.

Classroom Applications

Have girls take leadership roles in mathematics, such as serving as tutors in the classroom, or having them tutor students from lower grades. Assign girls to do a research paper on women in mathematics. In addition, there are lots of female mathematicians who would be pleased to come to a secondary school either to give a talk to a class on career options or to give a content talk to a mathematics class. A list of such persons can be obtained from the Mathematical Association of America or the American Mathematics Society. Further information can be found in the numerous publications from these organizations and commercial publishers. There are a number of books available from professional organizations and commercial publishers that address the issue of women in mathematics. The National Council of Teachers of Mathematics (1906 Association Drive, Reston, VA 20191-1593) offers the following suggestions: *Celebrating Women in Mathematics and Science, Multicultural and Gender Equity in the Mathematics Classroom, New Directions for Equity in Mathematics Education,* and *Reaching All Students With Mathematics.*

Precautions and Possible Pitfalls

Although one is usually delighted to find a proper role model for this purpose, it is imperative that the teacher interviews the potential speaker before the class presentation. This will give the teacher an idea about the visitor and help the teacher prepare the class for the ensuing presentation. In turn, the teacher can prepare the speaker for the class and any peculiarities (both favorable and unfavorable) about the class that it might be wise for the speaker to be aware of prior to the presentation. Keep in mind that parents may have stereotypes about women in mathematics that need to be overcome. Reach out to parents to let them know girls can be as successful as boys in mathematics, both in their schoolwork and in careers.

Source

Fennema, E., Wolleat, P. L., Pedro, J. D., & DeVaney Becker, A. (1981). Increasing women's participation in mathematics: An intervention study. *Journal for Research in Mathematics Education, 12*(1).

STRATEGY 76: Use different motivational strategies for girls and boys.

What the Research Says

When it comes to motivation, girls tend to be generalists while boys tend to be specialists. Interest, rather than intellect, often lies at the heart of the differences between boys and girls in mathematics. Girls tend to be interested in a wide range of subjects, while boys tend to concentrate their interests more narrowly. A study was conducted with 457 students; 338 students attended special mathematics- and science-oriented schools while 119 students attended regular schools but had excellent grades in mathematics, physics, and chemistry. At the beginning of a two-year study students were asked to rate their interest in later studying mathematics. Compared to boys, almost twice the percentage of girls showed interest in studying mathematics later. Several times over a period of two years teachers were asked to rank their students' abilities in mathematics. Ranking of girls became worse over the time.

Girls and boys were asked to rate how much they liked doing a variety of mathematical/physical and linguistic/literary tasks. Mathematical/physical tasks included finding variations of solutions to problems, solving especially difficult tasks, creating tasks by oneself, doing puzzles, and playing chess. Linguistic/literary tasks included making puns; following dialogues in literature, drama, or a radio play; having discussions with intellectuals; and finding contradictions or inconsistencies in texts. The results showed that girls are interested in a variety of areas and that they tend to concentrate their studying in all subjects rather than investing in one at the expense of the others, as boys tended to do. Over time, girls' interests expanded while boys' interests narrowed.

Teaching to the NCTM Standards

Although The Equity Principle for School Mathematics calls for "high expectations and strong support for all students,"[19] it does not support a one-size-fits-all approach. In the same manner that an experienced teacher will provide differentiated instruction to a class of students of varying abilities, teachers must utilize a variety of strategies to motivate different groups of students. Because sports are often used to motivate lessons in mathematics, teachers should be mindful that not all groups of students will find this inspirational. Effective teachers should develop a variety of motivational approaches that are not gender specific. Referencing art and music as outlined below may motivate both male and female students.

Classroom Applications

On average, girls often seem not to be as good in math as boys. This phenomenon does not happen because of girls' having less talent in math than boys. It is because of their greater interest in a wide range of topics. Consequently, girls will be more easily motivated if mathematical facts touch a wider range of subjects.

For example:

1. Relate the quadratic equation in general to that of the golden section, which can be shown as it exists in art or architecture.
2. Relate perspectivity in geometry to paintings, etchings, and drawings.
3. Connect mathematical structures with Bach's music.
4. Relate famous mathematicians like Pythagoras and Euclid to philosophy and/or history.

You might have students work on projects that correspond with their interests, and write papers or reports.

Precautions and Possible Pitfalls

Don't be disappointed if your efforts to motivate girls do not produce the desired effects. Continue to give girls the opportunity to demonstrate their abilities to achieve in mathematics.

Source

Käte Pollmer. Was behindert hochbegabte Mädchen, Erfolg im Mathematikunterricht zu erreichen? (What hinders highly talented girls from being successful in mathematics?). *Psychologie in Erziehung und Unterricht*, 38 Jg., (1991) S. 28–36.

STRATEGY 77: Take into consideration how students view successful teachers and how this differs for girls and boys.

What the Research Says

Girls are more critical of a teacher's appearance and behavior than are boys. Female teachers were evaluated more frequently than male teachers as unfair and as too soft. However, female teachers were regarded as less nervous, less disorderly, and more punctual than male teachers. Besides, female teachers did not make excessive demands on students and they smoked less. Female teachers were thoroughly evaluated by students in terms of their social behavior and their dress. Male teachers were evaluated more critically in terms of their politics and philosophy.

Participants were forty male and female teachers and their students from Grades 5 to 10.

The experiment consisted of four phases:

- *Phase 1.* Questionnaire that asked students for their view of the characteristics of effective teachers
- *Phase 2.* Observation of teachers' actions during lessons
- *Phase 3.* Analyses of teachers' personalities
- *Phase 4.* Procedures to improve or stabilize favorable teacher characteristics

The results showed that cheerfulness was the characteristic that was cited by students of all age groups. Most student views of the characteristics of effective teachers changed with the students' age. For example, being good at explaining facts was considered an effective characteristic by 34% of eighth graders, 41% of ninth graders, and 50% of tenth graders. The most frequent negative characteristics identified were being nervous, making excessive demands, being disorderly, being late, and smoking.

Thirteen positive characteristics of effective teachers were identified (see Figure 6.1):

Figure 6.1

1. Good methodology 2. Identifying the aim of a lesson 3. Dividing tasks into parts 4. Supplying or demanding summaries of a lesson 5. Tasks with a high degree of difficulty 6. Getting students to think automatically 7. Stimulating comments	8. Students actively involved in learning 9. Not making discouraging comments when students make mistakes 10. Focusing on the essential points 11. Giving students individual help or making individual demands 12. Giving varied assessments 13. Evaluating a student's personality from a positive perspective

Teaching to the NCTM Standards

The NCTM Professional Standards for Teaching and Learning state:

> Students' learning of mathematics is enhanced in a learning environment that is built as a community of people collaborating to make sense of mathematical ideas. It is a key function of the teacher to develop and nurture students' abilities to learn with and from others.[20]

Understanding specific characteristics that students view as important for effective teaching can help teachers shape their professional practice to take advantage of these influential factors. An examination of these traits reveals that most of them are obvious and support students' needs to feel important, respected, and valued as members of the learning community. The application below indicates that teachers can improve their learning environments by simply being pleasant and cheerful. This fosters an atmosphere where students feel comfortable taking intellectual risks, formulating conjectures, and justifying their answers in a nonthreatening environment. Teachers who display a sense of humor can make a class more enjoyable without sacrificing or compromising the instructional goals.

Classroom Applications

Teachers should be aware of the characteristics their students consider important. Female teachers are likely to find it harder to gain students' respect than male teachers. Because cheerfulness was considered important by students of all ages, teachers could benefit from making a deliberate attempt to be more cheerful, if they are not cheerful already. Mathematics teachers can influence the way in which students see them by delving into the field often referred to as *recreational mathematics.* This will garner greater teacher enthusiasm because of the novelty of the topics, and will then enable the teacher to stand out from the rest of the teachers.

Precautions and Possible Pitfalls

Being polite, always speaking gently to the students, encouraging them, and the like, is something that students expect from female teachers, but it alone does not work. A male teacher should not rely only on his professional knowledge. Although professional knowledge is important, there are other aspects of conducting a lesson that are important.

Source

Heinz Grassel. *Probleme und Ergebnisse von Untersuchungen der Lehrertätigkeit und Lehrerwirksamkeit* (Problems and results of investigations regarding teacher's activity and teacher's effectiveness). Studie des Wisssenschaftsbereichs Pädagogische Psychologie der Universität Rostock, 1968.

STRATEGY 78: Praise, encourage, and help your older students.

What the Research Says

Although older students often seem to avoid teachers' praise, research has shown that even older students care about being praised by the teacher. How do older and younger students compare to each other in how they feel about how their teachers treat them? A study was conducted with fourteen teachers: seven were primary school teachers who had a total of 194 students in Grades 1 through 4, and seven were secondary school teachers who had 167 students in Grades 5 through 10. Over a period of eight months students repeatedly filled out an anonymous questionnaire on their feelings about how teachers treat them. The questionnaire contained eighteen items that asked about how the teacher treated them. Questions addressed issues including teacher behaviors of friendliness, praise, encouragement, fairness, favoritism, display of anger, patience, and receptiveness to students' ideas. For each of the eighteen items students had to choose between whether teachers did or did not treat them in the ways described. For example, "The teacher praises me" or "The teacher doesn't praise me." The results showed that older and younger students had remarkably similar feelings. However, older students generally were more critical of their teachers than younger students. The most striking differences between older and younger students were in the following six areas:

a. The teacher praises me.
b. I like the ideas of the teacher.
c. I like taking an active part in the teacher's lessons.
d. The teacher encourages me.
e. The teacher helps me.
f. The teacher listens to what I say.

In the first five areas (a-e), a smaller percentage of older students were esatisfied with their teachers' behaviors than the younger students. In the last area (f), a higher percentage of older students were satisfied with their teachers' behaviors than the younger students.

Teaching to the NCTM Standards

The NCTM Teaching Principle states, "Effective mathematics teaching requires understanding what different students know and need to learn and then challenging and supporting them to do it well."[21] The research shows that as students get older, many teachers tend to present content with increasingly less concern for students' feelings and frustrations. Teachers should be on the lookout for frustration and know what to do when they encounter it. It is not always a bad thing! In the less proficient mathematics student, frustration may be a signal that the student is about to "shut down" and give up, instead of investing the effort to succeed. More talented mathematics students tend to channel frustration into taking on the challenge of solving a new, nonroutine problem. They see the reward and are willing to invest the time and effort to seek a solution. In either case, the teacher must recognize students' feelings so that appropriate action can be taken.

Classroom Applications

Especially for mathematics lessons, it is said that the communication between teachers and students seems too factual and unemotional. Older students in particular would benefit from more teacher praise, help, and encouragement. In math it is particularly important to be more sensitive when dealing with students who lack self-confidence in their mathematical abilities. Often teachers do not realize how their own behavior severely frustrates students. Sometimes a good relationship between students and a teacher suddenly changes and the teacher does not know why. Therefore it is important for teachers to find out about students' points of view, that is, teachers should see themselves through students' eyes. Investigate your students' feelings by using a questionnaire like the one described above, or develop one together with your students.

Precautions and Possible Pitfalls

Just listening to students and observing their behavior doesn't give teachers all the information they need to make intelligent decisions about how to treat students. Do not feel offended if you praise, help, or encourage students and they react in a disapproving way. Older, and especially mediocre, students often are embarrassed when treated like more ambitious students. Their embarrassment is the reason they react so aloofly when praised by a teacher, but under the surface they often feel pleased.

Source

Roswita von Hauff. Schüler geben ihren Lehrern Rückmeldung—Ein Rückmeldungsfragebogen, der von Grund- und Reallehrern sowie ihren Schülern entwickelt und erprobt wurde (Pupils give their teachers feedback—A feedback questionnaire developed and evaluated by teachers of primary schools and junior high schools). *Psychologie in Erziehung und Unterricht*, 29 Jg., (1982) S. 167–171.

STRATEGY 79: Does grade skipping hurt mathematically talented students socially and emotionally? Don't worry about accelerating talented students!

What the Research Says

Researchers conducted a long-term study of hundreds of male and female students who were identified as mathematically precocious youths in a talent search. The study began when students were twelve to fourteen years old and continued until they were twenty-three. The researchers examined the effects of the type and amount of acceleration students received on their social and emotional development. Two types of acceleration were examined: the number of grades skipped in school (grade acceleration) and the number of Advanced Placement and/or college courses enrolled in while in high school (subject matter acceleration). A questionnaire was administered to assess four areas of social and emotional development: self-esteem, self-acceptance/identity, social interaction, and locus of control (whether they internalize or externalize responsibility for their outcomes). Students were assessed when they were eighteen and again when they were twenty-three. The results indicated that acceleration did not affect students socially or emotionally. In addition, there were no differences between males and females.

Teaching to the NCTM Standards

The NCTM *Research Companion to Principles and Standards for School Mathematics* states that "students learn what they have an opportunity to learn."[22] Providing students opportunities to explore topics in depth and reach beyond the scope of the syllabus should be part of the professional practice of every classroom teacher. Students should be encouraged to pursue enrichment topics that extend the mathematical concepts that are presented in the classroom. The example provided below is one such way that acceleration can take place without simply teaching the material at a faster pace. This practice can help avoid the pitfall that often accompanies traditional acceleration: students who don't have the mathematical maturity to totally grasp and appreciate the material that is presented in the traditional accelerated fashion. Providing opportunities of this acceleration-enrichment model can also be a valuable way to provide a differentiated instructional model for teachers who have students of varying abilities in their class. While enrichment might not be appropriate for the entire class, students who are capable should be given the opportunity to explore enrichment topics.

Classroom Applications

Subject matter acceleration, when done creatively, can bear some interesting fruit. There is not much creativity in accelerating a student along an already established ladder of courses in mathematics. Yet, under such a scheme, special care must be taken to avoid some common pitfalls (see below). Again, to simply provide the student with the regular course work at a faster pace does little beyond keeping the student continuously challenged. A form of acceleration might be to move the student along more quickly, while at the same time offering him or her the opportunity to investigate the curriculum topics in a broader sense. For example, when studying the Pythagorean Theorem, rather than investigating the theorem and its traditional applications and then moving on to the next topic in the syllabus, it might be worthwhile to have the student investigate the Pythagorean Theorem's extensions beyond the right triangle to acute triangles ($a^2 + b^2 > c^2$) and to obtuse triangles ($a^2 + b^2 < c^2$). The student can also consider the extension of the Pythagorean Theorem to three dimensions, or to other fields such as number theory, where the nature of Pythagorean numbers is studied.

Precautions and Possible Pitfalls

One of the most serious problems attached to grade and subject matter acceleration is that of sending students through a set curriculum so fast that they don't truly get the full flavor of the subject matter, but rather just move along so as to finish as quickly as possible. Another point to consider is the problem of accelerating students along so quickly that they finish all that the school has to offer in mathematics and are then left without any further mathematics to study, yet still in high school finishing other courses. Such a "vacation" from mathematics will have a very deleterious effect on the future of a talented student, who could then be "lost" to the field of mathematics. This can be avoided by either providing these students with an independent study program under the tutelage of a teacher or enrolling them in an appropriate mathematics course at a nearby college. These considerations should be addressed before an acceleration program commences.

Source

Richardson, T. M., & Persson Benbow, C. (1990). Long-term effects of acceleration on the social-emotional adjustment of mathematically precocious youths. *Journal of Educational Psychology, 82*(3), 464–470.

Resource

What the Authors Say

Enriching Instruction

STRATEGY: Enrich your instruction before you look to accelerate the curriculum.

What the *Authors* Say

Recently, there has been a surge to move students along in the standard high school mathematics curriculum at a faster pace. In the course of this acceleration, a number of topics have been relegated to the back burner. In short, they were cut. Years ago, there were many useful techniques in algebra that were regularly taught but are not even mentioned today. One example is the teaching of *factoring*. Students were shown how to factor the sum and difference of two cubes, but today the practice is not mentioned. Every teacher owes it to his or her students to enrich their instruction. This can be done by expanding on a topic being taught, by extending the students' knowledge by showing how the topic being taught can relate to another topic out of the curriculum, or by bringing in historical aspects related to the topic.

Teaching to the NCTM Standards

The NCTM says enrichment is part of any good teaching performance. The Principles and Standards for School Mathematics state that "mathematics teachers generally are responsible for what happens in their own classrooms and can try to ensure that their classrooms support learning by all students . . . teachers must challenge and hold high expectations for all their students, not just those they believe are gifted."[1] It is a good professional practice to provide opportunities for all students to enrich their learning of mathematics by encouraging them to explore topics beyond the scope of the syllabus. The classroom application below contains some suggestions for student projects in mathematics.

Classroom Applications

There are a multitude of topics that can be used to enrich instruction. You can assign individual students to do a small report on a topic that relates to the material being taught. Listed below are some possible topics that students might use for such a project. This list is merely intended as a guide for generating additional topics.

Advanced Euclidean Geometry
Algebraic Fallacies
Algebraic Models
Algebraic Recreations
Analog Computer
Ancient Number Systems and Algorithms
Arithmetic Fallacies
Arithmetic Recreations
Bases Other Than Ten
Binary Computer
Boolean Algebra
Brocard Points
Calculating Shortcuts
Cavalieri's Theorem
Checking Arithmetic Operations
Conic Sections
Continued Fractions
Cryptography
Crystallography
Curves of Constant Breadth
Cylindrical Projections
Desargues' Theorem
Determinants
Diophantine Equations
Divisibility of Numbers
Duality
Dynamic Symmetry
Elementary Number Theory Applications
The Euler Line
Extension of Euler's Formula to *N* Dimensions
Extension of Pappus' Theorem
Fermat's Last Theorem
Fibonacci Numbers
Fields
Finite Differences
Finite Geometry
The Five Regular Polyhedra
Flexagons
The Four-Color Problem
The Fourth Dimension
Fractals
Game Theory
Gaussian Primes
Geodesics
Geometric Dissections: Tangrams
Geometric Fallacies
Geometric Models
Geometric Stereograms
Geometric Transformations
Geometry of Bubbles and Liquid Film
Geometry of Catenary
Geometry Constructions (Euclid)
Gergonne's Problem
The Golden Section

Graphical Representation of Complex Roots of Quadratic and Cubic Equations
Groups
Higher Algebra
Higher-Order Curves
Hyperbolic Functions
The Hyperbolic Paraboloid
Hypercomplex Numbers
Intuitive Geometric Recreations
Investigating the Cycloid
The Law of Growth
Linear Programming
Linkages
Lissajou's Figures
Lobachevskian Geometry
Logarithms of Negative and Complex Numbers
Logic
Magic Square Construction
Map Projections
Mascheroni's Constructions
Mathematics and Art
Mathematics and Music
Mathematics of Life Insurance
Matrices
Maximum-Minimum in Geometry
Means
Methods of Least Squares
The Metric System
Minimal Surfaces
Modulo Arithmetic in Algebra
Monte Carlo Method of Number Approximation
Multinomial Theorem
Napier's Rods
Networks
The Nine-Point Circle
Nomographs
The Number Pi, Phi, or *e*
Number Theory Proofs
Paper Folding
Partial Fractions
Pascal's Theorem
Perfect Numbers
Polygonal Numbers
Prime Numbers
Probability
Problem Solving in Algebra
Projective Geometry
Proofs of Algebraic Theorems
Properties of Pascal's Triangle
Pythagorean Theorem: Triples
Regular Polygons
The Regular Seventeen-Sided Polygon
Relativity and Mathematics
Riemannian Geometry
Solving Cubics and Quartics
Special Factoring
Spherical Triangles
The Spiral
Statistics
Steiner Constructions
Tesselations
Theory of Braids
Theory of Equations
Theory of Perspectives
Three-Dimensional Curves
The Three Famous Problems of Antiquity
Topology
Unsolved Problems
Vectors

You might also merely digress from a topic being taught. For example, when teaching probability you might mention the famous "birthday problem." When discussing concurrency in geometry, you might want to introduce Ceva's theorem, which makes many difficult theorems almost trivial to prove. We suggest you consider any of the following books, each of which will give you lots of ideas for enrichment:

Posamentier, A. S., & Hauptman H. A. (2001). *101 great ideas for introducing key concepts in mathematics.* Thousand Oaks, CA: Corwin Press.
Posamentier, A. S., Smith, B. S., & Stepelman, J. (2006). *Teaching secondary mathematics: Techniques and enrichment units.* Columbus, OH: Merrill/Prentice Hall.
Posamentier, A. S. (2003). *Math wonders to inspire students and teachers.* Alexandria, VA: ASCD.

Precautions and Possible Pitfalls

Whenever you embark on material that is not part of the curriculum and where students will essentially not be held responsible for learning the material, there is a tendency on the part of the students to stretch this activity out as long as possible. By keeping the teacher from moving along, figuring that if the teacher runs out of time and cannot cover a required topic, the students hope to have less material to study for when the next test comes along. So we urge you to do the enrichment with an eye towards a properly balanced time schedule. Don't let the enrichment material dominate the class activities. Remember that you are still responsible for covering the prescribed coursework. Experience will help you appropriately estimate the proper time allotment for such enrichment activities.

Epilogue

Now that you have read through the many research-based suggestions to improve your instruction, we want you to reflect on the larger picture of mathematics education today. Consider the following situation. A recent visit to a picture-framing shop highlighted a mathematical deficiency that seems to be common in our society. An inspection of the bill for framing two pictures, one four inches by twenty inches, and the other twelve inches by twelve inches, revealed that they cost the same. When questioned, the proprietor responded that the same amount of framing was used for the two pictures, and that the glass was figured on the basis of "united inches."

He was immediately asked what this unit of measure meant. He indicated that it was the sum of the length and the width; in this case each had twenty-four united inches, and so the cost was the same for the two pieces of glass. The merchant was asked if he believed the two frames required the same amount of glass. He wasn't sure, but assumed they did, since the two had the same number of united inches. A math teacher listening to this discussion chimed in to give him a quick lesson on rectangle area. The proprietor was amazed to discover that he had been charging the same amount for the two pieces of glass, when, in fact one's area (144 sq. in.) was almost twice that of the other (eighty sq. in.). This mathematical illiteracy is particularly alarming, especially in the context of our country's poor showing on a recent OECD-PISA study, where the United States came in twenty-eighth out of forty countries being compared on the mathematical achievement of fifteen-year-olds.

We have become complacent about achievement in mathematics. Adults more often than not take pride in their inability to have mastered school mathematics. Furthermore, when they are told that their children will need to master mathematics in school, they begin to question the reasons for such claims, especially when their children come home with math homework that looks unfamiliar to their parents. Over the years we have tried to convince others that there is power and beauty in mathematics. This is no easy task. We are often confronted with responses like, "I don't need to know arithmetic since I use the [ubiquitous] calculator." Or, "I don't even have to calculate the 'best buy' in the supermarket, since every item has its unit price indicated." Or "Even 'miles per gallon' need not be calculated, since my car's odometer does that for me." Some even ask, "Why teach mathematics at all?" Why don't they ever ask, "Why teach poetry, literature, music, art, or even science if one is not planning to pursue a career in those fields?" When was the last time an adult needed to *use* any of these subjects in everyday life?

We need to convince the general populace of the importance of mathematics. Simply saying, as many do, that today's students are involved with real-world applications in the classroom just doesn't cut it. Unfortunately, there are at least two problems with the real-world-math claims: First, the real world of students is often not what adults have chosen as the real world, and to be truly of the real world—rather than artificial models—is generally far too complicated for a school audience. There are times when parents do a "project" at home that involves mathematics or reasoning skills. Often these skills were developed as a result of school math instruction. Parents should involve their children in these projects, which might include setting up a birthday party, buying flooring or carpeting, or calculating expenses (i.e., budget). These would be actual real-world activities.

So where does that leave us? Do we merely stop teaching mathematics for the above-mentioned reasons, or do we try to demonstrate its purpose in other ways? Mathematics has manifested itself in the school curriculum in different ways at various times in our history. In the eighteenth century, students learned to "reckon," or do arithmetic, so that they would be able to do the necessary calculations required in trade or farming, for which they also needed some geometry. As time went on, the time available for math instruction increased, as did the material taught: algebra and trigonometry. More advanced instruction began to include some probability and statistics, although this was limited until the "number crunching" machines (i.e., calculators and computers) appeared. Today, we find ourselves with the dilemma of deciding exactly what and how mathematics should be taught. The advent of the computer has had a marked effect on the curriculum. Topics that used to be commonly taught are no longer needed, such as extracting the square root of a number or using logarithms to simplify complex calculations. In short, learning mathematics entails much more than obtaining tools to use in other fields. It is the subconscious acquisition of thinking and reasoning skills coupled with the more sophisticated way we view the physical world that leads the list of the many life enhancements that come with learning mathematics.

We, as members of the select New York State Math Standards Committee, were charged with preparing standards that would provide the necessary understandings and applications of arithmetic and geometry so that a proper transition could be made to a solid understanding and use of algebra, trigonometry, probability, and statistics. Today's youth needs a different facility with, and understanding of, numerical concepts than previous generations. The calculator and computer have reduced the need for one to be a lightning-fast calculator, yet despite today's technology, the need to understand number concepts and relationships has not diminished. Calculating a tip in a restaurant or balancing a checkbook still ought to be done easily.

For us to compete favorably in mathematics achievement in the world arena, we need to spend more time doing those things that school mathematics has done well for decades: provide appropriate reckoning skills, enable a reasonably sophisticated view of geometry for use in academics and beyond, and fortify students with the necessary tools of mathematics to pursue whatever academic endeavor they choose to study. The by-product of all of this will be a well-reasoning and able problem solver. Only if we do this in the context of motivating students with the beauty of mathematics (rather than telling them that what they are doing is for their "real-world" experiences, when they have difficulty accepting this) will our efforts bear fruit.

References

Chapter 1

1. Daniel Burke et al. (2004, March). *Relative pay and teacher retention in Miami-Dade County public schools: Summary of research.* Miami: Education Center Institute for Public Research.
2. The National Council of Teachers of Mathematics. (1991). Standard 5: Developing as a teacher of mathematics. In *Professional standards for teaching mathematics* (p. 160). Reston, VA: Author.
3. The National Council of Teachers of Mathematics. (2000). Principles for school mathematics: The learning principle. In *Principles and standards for school mathematics* (p. 20). Reston, VA: Author.
4. The National Council of Teachers of Mathematics. (2003). Communication and language: How is mathematics taught and learned in classroom communication? In *A research companion to principles and standards for school mathematics* (p. 244). Reston, VA: Author.
5. The National Council of Teachers of Mathematics. (2003). Communication and language: How is mathematics taught and learned in classroom communication? In *A research companion to principles and standards for school mathematics* (p. 244). Reston, VA: Author.
6. The National Council of Teachers of Mathematics. (1991). Standard 5: Learning environment. In *Professional standards for teaching mathematics* (p. 57). Reston, VA: Author.
7. The National Council of Teachers of Mathematics. (1991). Standard 2: The teacher's role in discourse. In *Professional standards for teaching mathematics* (p. 35). Reston, VA: Author.
8. The National Council of Teachers of Mathematics. (1991). Standard 5: Learning environment. In *Professional standards for teaching mathematics* (p. 58). Reston, VA: Author.
9. The National Council of Teachers of Mathematics. (2000). Principles for school mathematics: The learning principle. In *Principles and standards for school mathematics* (p. 21). Reston, VA: Author.
10. The National Council of Teachers of Mathematics. (2000). Principles for school mathematics: The learning principle. In *Principles and standards for school mathematics* (p. 21). Reston, VA: Author.
11. The National Council of Teachers of Mathematics. (2000). Principles for school mathematics: The learning principle. In *Principles and standards for school mathematics* (p. 21). Reston, VA: Author.

12. The National Council of Teachers of Mathematics. (1991). Analysis. In *Professional standards for teaching mathematics* (p. 62). Reston, VA: Author.
13. The National Council of Teachers of Mathematics. (1991). Analysis. In *Professional standards for teaching mathematics* (p. 62). Reston, VA: Author.
14. The National Council of Teachers of Mathematics. (1992). Overview: Learning and teaching with understanding. In *The handbook of research on mathematics teaching and learning* (p. 74). Reston, VA: Author.
15. The National Council of Teachers of Mathematics. (1992). Overview: Learning and teaching with understanding. In *The handbook of research on mathematics teaching and learning* (p. 74). Reston, VA: Author.
16. The National Council of Teachers of Mathematics. (1992). Overview: Learning and teaching with understanding. In *The handbook of research on mathematics teaching and learning* (p. 75). Reston, VA: Author.
17. The National Council of Teachers of Mathematics. (1991). Standard 5: Learning environment. In *Professional standards for teaching mathematics* (pp. 57–58). Reston, VA: Author.
18. The National Council of Teachers of Mathematics. (1991). Standard 5: Learning environment. In *Professional standards for teaching mathematics* (p. 57). Reston, VA: Author.
19. The National Council of Teachers of Mathematics. (1992). Mathematics teaching: Grouping for instruction in mathematics: A call for programmatic research on small-group processes. In *The handbook of research on mathematics teaching and learning* (p. 181). Reston, VA: Author.
20. The National Council of Teachers of Mathematics. (1992). Mathematics teaching: Grouping for instruction in mathematics: A call for programmatic research on small-group processes. In *The handbook of research on mathematics teaching and learning* (p. 181). Reston, VA: Author.
21. The National Council of Teachers of Mathematics. (1992). Mathematics teaching: Grouping for instruction in mathematics: A call for programmatic research on small-group processes. In *The handbook of research on mathematics teaching and learning* (p. 181). Reston, VA: Author.
22. The National Council of Teachers of Mathematics. (2000). Standards for school mathematics: Communication. In *Principles and standards for school mathematics* (p. 60). Reston, VA: Author.
23. The National Council of Teachers of Mathematics. (1991). Standard 5: Learning environment. In *Professional standards for teaching mathematics* (p. 61). Reston, VA: Author.
24. The National Council of Teachers of Mathematics. (1991). Evaluation of teaching: Standard 5: Mathematics as problem solving, reasoning, and communication. In *Professional standards for teaching mathematics* (p. 95). Reston, VA: Author.
25. The National Council of Teachers of Mathematics. (1991). Evaluation of teaching: Standard 5: Mathematics as problem solving, reasoning, and communication. In *Professional standards for teaching mathematics* (p. 96). Reston, VA: Author.
26. The National Council of Teachers of Mathematics. (1992). Critical issues: Race, ethnicity, social class, language, and achievement in mathematics. In *The handbook of research on mathematics teaching and learning* (p. 652). Reston, VA: Author.

27. The National Council of Teachers of Mathematics. (1992). Critical issues: Race, ethnicity, social class, language, and achievement in mathematics. In *The handbook of research on mathematics teaching and learning* (p. 652). Reston, VA: Author.

Chapter 2

1. The National Council of Teachers of Mathematics. (2000). Principles for school mathematics: The teaching principle. In *Principles and standards for school mathematics* (p. 16). Reston, VA: Author.
2. The National Council of Teachers of Mathematics. (2000). Standards for school mathematics: Connections. In *Principles and standards for school mathematics* (p. 64). Reston, VA: Author.
3. The National Council of Teachers of Mathematics. (2000). Standards for school mathematics: Connections. In *Principles and standards for school mathematics* (p. 64). Reston, VA: Author.
4. The National Council of Teachers of Mathematics. (2000). Principles for school mathematics: The teaching principle. In *Principles and standards for school mathematics* (p. 18). Reston, VA: Author.
5. The National Council of Teachers of Mathematics. (2000). Standards for school mathematics: The communication standard. In *Principles and standards for school mathematics* (p. 60). Reston, VA: Author.
6. The National Council of Teachers of Mathematics. (2000). Standards for school mathematics: The communication standard. In *Principles and standards for school mathematics* (p. 60). Reston, VA: Author.
7. The National Council of Teachers of Mathematics. (1991). Standard 6: Promoting mathematical disposition. In *Professional standards for teaching mathematics* (p. 104). Reston, VA: Author.
8. The National Council of Teachers of Mathematics. (2003). Classroom and large-scale assessment: Issues in classroom assessment. In *A research companion to principles and standards for school mathematics* (p. 55). Reston, VA: Author.
9. The National Council of Teachers of Mathematics. (2003). Communication and language: How is mathematics taught and learned in classroom communication? In *A research companion to principles and standards for school mathematics* (p. 244). Reston, VA: Author.
10. Ibid., p. i.
11. Ibid., p. i.
12. The National Council of Teachers of Mathematics. (2003). Communication and language: How is mathematics taught and learned in classroom communication? In *A research companion to principles and standards for school mathematics* (p. 245). Reston, VA: Author.
13. Ibid., p. iv.
14. The National Council of Teachers of Mathematics. (2000). Standards for Grades 9–12: Reasoning and proof standards for Grades 9–12. In *Principles and standards for school mathematics* (p. 342). Reston, VA: Author.
15. The National Council of Teachers of Mathematics. (2000). Standards for school mathematics: Connections. In *Principles and standards for school mathematics* (p. 64). Reston, VA: Author.

16. The National Council of Teachers of Mathematics. (2003). Communication and language: Teaching mathematical communication to all students. In *A research companion to principles and standards for school mathematics* (p. 246). Reston, VA: Author.
17. The National Council of Teachers of Mathematics. (2000). Principles for school mathematics: The learning principle. In *Principles and standards for school mathematics* (p. 21). Reston, VA: Author.
18. The National Council of Teachers of Mathematics. (1992). Critical issues: Ethnomathematics and everyday cognition. In *The handbook of research on mathematics teaching and learning* (p. 557). Reston, VA: Author.
19. The National Council of Teachers of Mathematics. (1991). Standard 2: The teacher's role in discourse. In *Professional standards for teaching mathematics* (p. 35). Reston, VA: Author.
20. The National Council of Teachers of Mathematics. (1995). Purpose: Monitoring students' progress. In *Assessment standards for school mathematics* (p. 29). Reston, VA: Author.
21. The National Council of Teachers of Mathematics. (1992). Learning from instruction: Geometry and spatial reasoning. In *The handbook of research on mathematics teaching and learning* (p. 454). Reston, VA: Author.

Chapter 3

1. National Council of Teachers of Mathematics. (2000). Appendix: Table of standards and expectations. In *Principles and standards for school mathematics.* Reston, VA: Author.
2. The National Council of Teachers of Mathematics. (2000). Principles for school mathematics: The learning principle. In *Principles and standards for school mathematics* (p. 20). Reston, VA: Author.
3. The National Council of Teachers of Mathematics. (2003). Representation in school mathematics: Learning to graph and graphing to learn. In *A research companion to principles and standards for school mathematics* (p. 260). Reston, VA: Author.
4. The National Council of Teachers of Mathematics. (2000). Standards for school mathematics: Problem solving. In *Principles and standards for school mathematics* (p. 334). Reston, VA: Author.
5. The National Council of Teachers of Mathematics. (2003). Communication and language: What is taught and learned in mathematical communication? In *A research companion to principles and standards for school mathematics* (p. 242). Reston, VA: Author.
6. The National Council of Teachers of Mathematics. (2000). Standards for Grades 9–12: Communication standards for Grades 9–12. In *Principles and standards for school mathematics* (p. 348). Reston, VA: Author.
7. The National Council of Teachers of Mathematics. (1991). Standard 3: Students' role in discourse. In *Professional standards for teaching mathematics* (p. 45). Reston, VA: Author.
8. The National Council of Teachers of Mathematics. (1991). Standard 3: Students' role in discourse. In *Professional standards for teaching mathematics* (p. 45). Reston, VA: Author.

9. The National Council of Teachers of Mathematics. (2000). Standards for school mathematics: Problem solving. In *Principles and standards for school mathematics* (p. 52). Reston, VA: Author.
10. The National Council of Teachers of Mathematics. (1989). Standard 3: Mathematics and reasoning. In *Curriculum and evaluation standards for school mathematics* (p. 143). Reston, VA: Author.
11. The National Council of Teachers of Mathematics. (1989). Standard 4: Mathematics connections. In *Curriculum and evaluation standards for school mathematics* (p. 146). Reston, VA: Author.
12. The National Council of Teachers of Mathematics. (2000). Standards for school mathematics: The communication standard. In *Principles and standards for school mathematics* (p. 60). Reston, VA: Author.
13. The National Council of Teachers of Mathematics. (2000). Principles for school mathematics: The learning principle. In *Principles and standards for school mathematics* (p. 21). Reston, VA: Author.
14. The National Council of Teachers of Mathematics. (2003). What research says about the NCTM standards: Student learning with traditional curricula and pedagogy. In *A research companion to principles and standards for school mathematics* (pp. 10–11). Reston, VA: Author. Fey, J. (1979). Mathematics teaching today: Perspectives from three national surveys. *Mathematics Teacher, 72,* 490–504.
15. The National Council of Teachers of Mathematics. (2003). What research says about the NCTM Standards: Student learning with traditional curricula and pedagogy. In *A research companion to principles and standards for school mathematics* (p. 11). Reston, VA: Author. Stigler, J. W., & Hiebert, J. (1997). Understanding and improving classroom mathematics instruction: An overview of the TIMSS video study. *Phi Delta Kappan, 79*(1), 14–21.
16. The National Council of Teachers of Mathematics. (2003). What research says about the NCTM Standards: Student learning with traditional curricula and pedagogy. In *A research companion to principles and standards for school mathematics* (p. 11). Reston, VA: Author.
17. The National Council of Teachers of Mathematics. (2000). Standards for school mathematics: Connections. In *Principles and standards for school mathematics* (p. 64). Reston, VA: Author.
18. The National Council of Teachers of Mathematics. (1991). Standard 4: Tools for enhancing discourse. In *Professional standards for teaching mathematics* (p. 52). Reston, VA: Author.
19. The National Council of Teachers of Mathematics. (1991). Standard 4: Tools for enhancing discourse. In *Professional standards for teaching mathematics* (p. 52). Reston, VA: Author.
20. The National Council of Teachers of Mathematics. (1991). Standard 3: Students' role in discourse. In *Professional standards for teaching mathematics* (p. 45). Reston, VA: Author.

Chapter 4

1. The National Council of Teachers of Mathematics. (1995). The learning standard. In *Assessment standards for school mathematics* (p. 14). Reston, VA: Author.

2. The National Council of Teachers of Mathematics. (1995). The learning standard. In *Assessment standards for school mathematics* (p. 14). Reston, VA: Author.
3. The National Council of Teachers of Mathematics. (1995). The learning standard. In *Assessment standards for school mathematics* (p. 14). Reston, VA: Author.
4. The National Council of Teachers of Mathematics. (1995). The learning standard. In *Assessment standards for school mathematics* (p. 14). Reston, VA: Author.
5. The National Council of Teachers of Mathematics. (1991). Standard 2: The teacher's role in discourse. In *Professional standards for teaching mathematics* (p. 35). Reston, VA: Author.
6. The National Council of Teachers of Mathematics. (1991). Standard 2: The teacher's role in discourse. In *Professional standards for teaching mathematics* (p. 45). Reston, VA: Author.
7. The National Council of Teachers of Mathematics. (2000). Principles for school mathematics: The learning principle. In *Principles and standards for school mathematics* (p. 20). Reston, VA: Author.
8. The National Council of Teachers of Mathematics. (1992). Overview: Learning and teaching with understanding. In *The handbook of research on mathematics teaching and learning* (p. 67). Reston, VA: Author.
9. The National Council of Teachers of Mathematics. (2000). Principles for school mathematics: The teaching principle. In *Principles and standards for school mathematics* (p. 16). Reston, VA: Author.
10. The National Council of Teachers of Mathematics. (2000). Principles for school mathematics: The learning principle. In *Principles and standards for school mathematics* (p. 20). Reston, VA: Author.
11. The National Council of Teachers of Mathematics. (2000). Principles for school mathematics: The assessment principle. In *Principles and standards for school mathematics* (p. 22). Reston, VA: Author.
12. The National Council of Teachers of Mathematics. (2000). Principles for school mathematics: The assessment principle. In *Principles and standards for school mathematics* (p. 22). Reston, VA: Author.
13. The National Council of Teachers of Mathematics. (2003). Classroom and large-scale assessment: Issues in classroom assessment. In *A research companion to principles and standards for school mathematics* (p. 58). Reston, VA: Author.
14. The National Council of Teachers of Mathematics. (2003). Classroom and large-scale assessment: Issues in classroom assessment. In *A research companion to principles and standards for school mathematics* (p. 58). Reston, VA: Author.
15. The National Council of Teachers of Mathematics. (1995). The learning standard. In *Assessment standards for school mathematics* (p. 13). Reston, VA: Author.
16. The National Council of Teachers of Mathematics. (1995). The learning standard. In *Assessment standards for school mathematics* (p. 14). Reston, VA: Author.
17. The National Council of Teachers of Mathematics. (1991). Standard 5: Learning environment. In *Professional standards for teaching mathematics* (p. 57). Reston, VA: Author.

Chapter 5

1. The National Council of Teachers of Mathematics. (2000). Standards for school mathematics: Communication. In *Principles and standards for school mathematics* (p. 60). Reston, VA: Author.
2. The National Council of Teachers of Mathematics. (1991). Standard 6: Analysis of teaching and learning. In *Professional standards for teaching mathematics* (p. 63). Reston, VA: Author.
3. The National Council of Teachers of Mathematics. (2000). Standards for school mathematics: Problem solving. In *Principles and standards for school mathematics* (p. 52). Reston, VA: Author.
4. The National Council of Teachers of Mathematics. (2000). Principles for school mathematics: The learning principle. In *Principles and standards for school mathematics* (p. 20). Reston, VA: Author.
5. The National Council of Teachers of Mathematics. (2000). Standards for school mathematics: Problem solving. In *Principles and standards for school mathematics* (p. 52). Reston, VA: Author.
6. The National Council of Teachers of Mathematics. (2000). Standards for Grades 9–12: Problem solving standards. In *Principles and standards for school mathematics* (p. 334). Reston, VA: Author.
7. The National Council of Teachers of Mathematics. (2000). *Principles and standards for school mathematics* (p. 187). Reston, VA: Author.
8. The National Council of Teachers of Mathematics. (2000). Principles for school mathematics: The learning principle. In *Principles and standards for school mathematics* (p. 21). Reston, VA: Author.
9. The National Council of Teachers of Mathematics. (2000). Standards for school mathematics: Problem solving. In *Principles and standards for school mathematics* (p. 52). Reston, VA: Author.
10. The National Council of Teachers of Mathematics. (1991). Standard 2: The teacher's role in discourse. In *Professional standards for teaching mathematics* (p. 35). Reston, VA: Author.
11. The National Council of Teachers of Mathematics. (1991). Standard 2: The teacher's role in discourse. In *Professional standards for teaching mathematics* (p. 36). Reston, VA: Author.
12. The National Council of Teachers of Mathematics. (2000). Principles for school mathematics: The learning principle. In *Principles and standards for school mathematics* (p. 20). Reston, VA: Author.

Chapter 6

1. The National Council of Teachers of Mathematics. (1992). Critical issues: Ethnomathematics and everyday cognition. In *The handbook of research on mathematics teaching and learning* (p. 557). Reston, VA: Author.
2. The National Council of Teachers of Mathematics. (1992). Critical issues: Ethnomathematics and everyday cognition. In *The handbook of research on mathematics teaching and learning* (p. 557). Reston, VA: Author.

3. The National Council of Teachers of Mathematics. (1992). Critical issues: Ethnomathematics and everyday cognition. In *The handbook of research on mathematics teaching and learning* (p. 558). Reston, VA: Author. Stigler, J. W., & Baranes, R. (1988). Culture and mathematics learning. In E. Z. Rothkopf (Ed.), *Review of research in education* (Vol. 15, pp. 253–306). Washington, DC: American Educational Research Association.
4. The National Council of Teachers of Mathematics. (1992). Critical issues: Ethnomathematics and everyday cognition. In *The handbook of research on mathematics teaching and learning* (p. 558). Reston, VA: Author.
5. The National Council of Teachers of Mathematics. (1992). Critical issues: Ethnomathematics and everyday cognition. In *The handbook of research on mathematics teaching and learning* (p. 558). Reston, VA: Author.
6. The National Council of Teachers of Mathematics. (1992). Critical issues: Mathematics and gender: Changing perspectives. In *The handbook of research on mathematics teaching and learning* (p. 613). Reston, VA: Author.
7. The National Council of Teachers of Mathematics. (1992). Critical issues: Mathematics and gender: Changing perspectives. In *The handbook of research on mathematics teaching and learning* (p. 613). Reston, VA: Author.
8. The National Council of Teachers of Mathematics. (1992). Critical issues: Mathematics and gender: Changing perspectives. In *The handbook of research on mathematics teaching and learning* (p. 613). Reston, VA: Author.
9. The National Council of Teachers of Mathematics. (1992). Critical issues: Mathematics and gender: Changing perspectives. In *The handbook of research on mathematics teaching and learning* (p. 613). Reston, VA: Author.
10. The National Council of Teachers of Mathematics. (1992). Critical issues: Mathematics and gender: Changing perspectives. In *The handbook of research on mathematics teaching and learning* (p. 613). Reston, VA: Author.
11. The National Council of Teachers of Mathematics. (2000). Principles for school mathematics: The assessment principle. In *Principles and standards for school mathematics* (p. 22). Reston, VA: Author.
12. The National Council of Teachers of Mathematics. (2000). Principles for school mathematics: The equity principle. In *Principles and standards for school mathematics* (p. 12). Reston, VA: Author.
13. The National Council of Teachers of Mathematics. (1992). Critical issues: Mathematics and gender: Changing perspectives. In *The handbook of research on mathematics teaching and learning* (p. 610). Reston, VA: Author.
14. The National Council of Teachers of Mathematics. (1992). Critical issues: Mathematics and gender: Changing perspectives. In *The handbook of research on mathematics teaching and learning* (p. 611). Reston, VA: Author.
15. The National Council of Teachers of Mathematics. (2000). Principles for school mathematics: The equity principle. In *Principles and standards for school mathematics* (p. 12). Reston, VA: Author.
16. The National Council of Teachers of Mathematics. (1992). Critical issues: Mathematics and gender: Changing perspectives. In *The handbook of research on mathematics teaching and learning* (p. 614). Reston, VA: Author. Fennema, E., & Sherman, J. (1977). Sex-related differences in mathematics achievement, spatial visualization, and sociocultural factors. *American Educational Research Journal, 14,* 51–71. Fennema, E., & Sherman, J. A. (1978). Mathematics achievement and related factors: A further study. *Journal for Research in Mathematics Education, 9,* 189–203.

17. The National Council of Teachers of Mathematics. (1992). Critical issues: Mathematics and gender: Changing perspectives. In *The handbook of research on mathematics teaching and learning* (p. 614). Reston, VA: Author. Joffe, L., & Foxman, D. (1986). Attitudes and sex differences: Some APU findings. In L. Burton (Ed.), *Girls into maths can go* (pp. 38–50). London: Holt, Rinehart & Winston.
18. The National Council of Teachers of Mathematics. (1992). Critical issues: Mathematics and gender: Changing perspectives. In *The handbook of research on mathematics teaching and learning* (p. 614). Reston, VA: Author. Joffe, L., & Foxman, D. (1988). *Attitudes and gender differences: Mathematics at age 11 and 15.* Windsor, Berkshire, UK: NFER-Nelson.
19. The National Council of Teachers of Mathematics. (2000). Principles for school mathematics: The equity principle. In *Principles and standards for school mathematics* (p. 12). Reston, VA: Author.
20. The National Council of Teachers of Mathematics. (1991). Standard 5: Learning environment. In *Professional standards for teaching mathematics* (p. 58). Reston, VA: Author.
21. The National Council of Teachers of Mathematics. (2000). Principles for school mathematics: The teaching principle. In *Principles and standards for school mathematics* (p. 16). Reston, VA: Author.
22. The National Council of Teachers of Mathematics. (2003). What research says about the NCTM Standards: Baseline conclusion: Students learn what they have an opportunity to learn. In *A research companion to principles and standards for school mathematics* (p. 10). Reston, VA: Author.

Resource

1. The National Council of Teachers of Mathematics. (2000). *Principles and standards for school mathematics* (p. 375). Reston, VA: Author.

Index

The Corwin Press logo—a raven striding across an open book—represents the union of courage and learning. Corwin Press is committed to improving education for all learners by publishing books and other professional development resources for those serving the field of PreK–12 education. By providing practical, hands-on materials, Corwin Press continues to carry out the promise of its motto: **"Helping Educators Do Their Work Better."**